BLACK HOLES
and
the Universe

IGOR NOVIKOV
Space Research Institute, Moscow

TRANSLATED BY VITALY KISIN

CAMBRIDGE
UNIVERSITY PRESS

Published by the Press Syndicate of the University of Cambridge
The Pitt Building, Trumpington Street, Cambridge CB2 1RP
40 West 20th Street, New York, NY 10011–4211, USA
10 Stamford Road, Oakleigh, Victoria 3166, Australia

First published 1990
Reprinted 1990, 1992
Canto edition 1995

Printed in Great Britain at the University Press, Cambridge

British Library cataloguing in publication data
Novikov, Igor
Black holes and the universe.
1. Black holes
I. Title
523

Library of Congress cataloguing in publication data
Novikov, I. D. (Igor Dmitrievich)
[Chernye dyry 1 Vselennaia. English]
Black holes and the universe / Igor Novikov; translated by Vitaly Kisin.
p.   cm.
Translation of: Chernye dyry i Vselennaia
ISBN 0 521 36657 7 – ISBN 0 521 36683 6 (paperback)
1. Black holes (Astronomy) 2. Cosmology. I. Title
QB843.B55N6713 1990
523.8′875–dc20   89-38038   CIP

ISBN 0 521 36657 7 hardback
ISBN 0 521 55870 0 paperback

Contents

Preface to the English Edition

The superstrong gravitational field is the protagonist of this book. This gravitation is the power that warps space and time into a funnel and generates a black hole when a cosmic body undergoes catastrophic collapse. This superstrong gravitation reigns in the Universe, controlling the motion of infinitely large masses.

The book describes natural phenomena caused by superstrong gravitation but perceived as nothing short of miracles, but it also explains how these miracles are studied and understood.

It is a special pleasure for me that the English translation of this book is published by the Cambridge University Press. Cambridge was the cradle of the science of gravitation. The great Newton, whose name was given to the law that rules the motion of worlds, worked in Cambridge. It was to Cambridge that the English geologist John Michell sent a letter addressed to the famous physicist Henry Cavendish where he predicted the existence of the most peculiar objects that we now call black holes. It was in Cambridge that pulsars were discovered in the middle of this century. (Pulsars are stars, about as massive as our Sun, that are compressed by fantastically strong gravity to a size of about ten kilometers in diameter.) Now, as before, Cambridge is the place of research for outstanding scientists, such as Stephen Hawking, Martin Rees, and others who study the mysteries of evaporation of black holes and the puzzle of the origin of the enigmatic pattern created by the large-scale structure of the Universe. For a Soviet physicist and astrophysicist, contacts with British colleagues, and especially those at Cambridge, were always exceptionally fruitful. Each of my visits to Cambridge created a feeling of inspiration and enthusiasm.

I am deeply grateful to Cambridge University Press and also to my colleague in Moscow, Vitaly Kisin, who translated this book into English, for doing everything to make this popular science text reach the English reader.

Moscow   *Igor Novikov*

Introduction

This book tells the story of the recent discoveries in astrophysics: black holes and the beginning of the expansion of the Universe; it also outlines what the future of the Universe may be.

No doubt, every reader has heard or read about black holes. They are frequently mentioned in television and radio programs, in magazines and all sorts of books, from scientific treatises to fiction and even books for children. What makes them so popular?

The point is that black holes are objects whose properties are absolutely fantastic.

Of all the conceptions of the human mind from unicorns to gargoyles to the hydrogen bomb perhaps the most fantastic is the black hole: a hole in space with a definite edge over which anything can fall and nothing can escape; a hole with a gravitational field so strong that even light is caught and held in its grip; a hole that curves space and warps time. Like the unicorn and the gargoyle the black hole seems much more at home in science fiction or in ancient myth than in real universe. Nevertheless, the laws of modern physics virtually demand that black holes exist. In our galaxy alone there may be millions of them.

This characteristic was given to black holes by the American physicist Kip Thorn.

One has to add that the properties of space and time are changed inside the black hole in a most puzzling manner: they get coiled into what resembles a funnel, with a boundary deep inside it beyond which time and space break down into quanta. The black hole hides in its depths, over the edge of this 'gravitational abyss' 'of no escape', astonishing physical processes and manifests new laws of nature.

Black holes are the most grandiose energy sources in the Universe. It appears likely that what we observe in remote quasars and in the exploding nuclei of galaxies are manifestations of black holes. Black holes are also created when massive stars die. It cannot be excluded that some day mankind will be able to utilize black holes as energy sources.

Spectacular discoveries were reported by astrophysicists studying the 'Big Bang', as the observed expansion of the Universe came to be known. When did the expansion begin and why? What were the events immediately after the Big Bang started? How and for what reason did matter appear in the Universe? Why did galaxies form? Is the Universe infinite? What does the future hold for the Universe? These are the problems discussed in this book.

A natural question arises: why do we describe black holes and the Universe as a whole within a single book? What do they have in common? The answer can be given by a very short phrase. From the standpoint of physics, the common feature is the superstrong gravitational field. Today both black holes and the Universe as a whole are studied in the framework of relativistic astrophysics: the branch of astrophysics that analyzes the processes in which the gravitational field is so strong that they accelerate bodies to nearly the velocity of light, that is, to the maximum velocity allowed in nature. No other object has gravitational fields reaching this high level. But we can also say that black holes and the Universe have another common feature: they are mysterious, puzzling, unusual. Their properties differ drastically not only from the properties of things that surround us daily but also from the properties of numerous physical objects and heavenly bodies that quite frequently are far from trivial themselves.

The more mysterious a puzzle, the more profound a problem, and the more intense becomes the interest displayed by specialists and laymen alike. Albert Einstein, the creator of general relativity theory, wrote: 'The most refined and profound emotional experience that a man may be honoured with is the feeling of mystery.' It is unlikely that black holes have any competition in the Universe as far as mysteriousness goes.

Finally, we can point to one more factor, a significant one, that necessitates a joint story. The Universe is known to be expanding. The density of matter in the Universe in the very remote past was colossal. Energies of particles and interaction forces between them were unimaginably high. These were the conditions under which very unusual laws of nature were reigning and even the properties of space and time changed dramatically. This state was said to be singular: it is known as 'the singularity'. On the other hand, a singularity develops with inevitability inside a black hole. For this reason, a black hole is sometimes regarded as a laboratory in which the past of our Universe is modelled on a miniature scale. It is no wonder, therefore, that quite frequently the same scientists study the evolution of both the Universe and black holes (the author of this book is one of them).

At the end of the thirteenth century people used the maxim: 'Avoid two explanations where one suffices'; nevertheless, we have chosen to give several reasons that convinced us that the book should combine the descriptions of these, seemingly so dissimilar, problems.

This book is aimed at those who love profound scientific mysteries. It describes a number of specific peculiarities and properties of black holes and of the Universe, and tells the story of how some problems were solved and how new ones arose. I have made no attempt to give an exhaustive presentation of all details: this would be meaningless and unrealistic for a book of this type. For the same reasons, I have mentioned dates and names only occasionally, mostly in connection with my personal impressions and reminiscences. Rather, I have intended to create, using separate strokes, an image of the grandiose problems solved or only tackled by astrophysicists.

Part I
Black holes

1: *What is a black hole?*

Stars that the world cannot see

A black hole is a creation of gravitation. It is thus logical to start the account of the discovery of black holes at the time of Isaac Newton, who discovered the law of universal gravitation. This law determines the force acting on absolutely everything. All other types of physical interaction are connected with concrete properties of matter. For instance, the electric field acts only on charged bodies while neutral ones are perfectly indifferent to it. Gravitation is the only interaction that reigns supreme in nature. It acts on everything: on small-mass and large-mass particles (in precisely the same way if initial conditions are identical), even on light. Newton hypothesized that light is attracted to massive bodies. It is with this understanding of the fact that light is also subject to gravitational force that the early history of black holes begins, the history of prediction of their astonishing properties.

One of the first scientists to come up with the prediction was the famous French mathematician and astronomer Pierre Laplace, an illustrious personality in the history of science. First and foremost, he was the author of the enormous five-volume *Treatise on Celestial Mechanics*. This work, written and published over a period from 1798 to 1825, presented the classical theory of motion of bodies in the Solar System, based solely on the law of universal gravitation.

Before that, some observed peculiarities in the motion of planets, the Moon, and other bodies of the Solar System could not be completely understood. It even seemed that they contradicted Newton's law. Elegant mathematical analysis performed by Laplace proved that all these finer points are caused by the mutual attraction of celestial bodies, by the effects of the gravitation of planets on the motion of other planets. He asserted that the only force reigning in the skies is the force of gravitation. In the preface to his *Treatise*, Laplace wrote that from the most general standpoint, astronomy is the grand problem in mechanics. Note that he was the first to suggest the term 'celestial mechanics', now firmly rooted in the language of science.

Laplace was also one of the first to realize the need in a historical approach for explaining the properties of systems of celestial bodies. He advanced, after Kant, a hypothesis of the formation of the Solar System from the originally rarefied matter.

The principal idea of Laplace's hypothesis of the condensation of the Sun and the planets from a gas nebula still forms the foundation of currently developed theories of the origin of the Solar system.

Much was written in the literature and in textbooks about these achievements, as well as about Laplace's proud reply to Napoleon's question about no reference to God in *Celestial Mechanics*: 'I did not need this hypothesis.'

The point that was little known until recently was that Laplace predicted the existence of invisible stars.

The prediction was made in his monograph *Exposition of the System of the World* published in 1795. In this book – which would be considered 'popular' by today's standards – the famous mathematician chose not to resort to either formulas or drawings. Laplace's profound conviction that gravitation affects light in the same way that it acts on other objects led him to the following spectacular conclusion:

A luminous star, of the same density as the Earth, and whose diameter should be two hundred and fifty times larger than that of the Sun, would not, in consequence of its attraction, allow any of its rays to arrive at us; it is therefore possible that the largest luminous bodies in the universe may, through this cause, be invisible.

The book did not give the proof of this proposition. It was published several years later.

How did Laplace reason? Using Newton's theory of gravitation, he calculated what we now call the escape velocity. This is the velocity that we have to impart to a body (any body) in order for it to break away from the gravitational pull of a star or planet and fly away for ever into space. If the initial velocity of the body is less than the escape velocity, gravitational forces will decelerate its motion, stop the body, and then make it fall back on the gravitating center. Everybody is supposed to know in this age of space flight that the escape velocity on the surface of the Earth is 11 kilometers per second. The greater the mass and the smaller the radius of a celestial body the greater is the escape velocity on the surface. This is clear: as mass increases, gravitation is intensified, and as the distance from the center increases, gravitation is weakened.

The escape velocity equals 2.4 km/s on the surface of the Moon, 61 km/s on Jupiter, 620 km/s on the Sun, and reaches one half of the velocity of light, that is, 150 000 km/s, on the surface of so-called neutron stars whose mass is close to one solar mass but whose radius is only about ten kilometers.

Laplace argued like this: take a celestial body on whose surface the escape velocity exceeds the velocity of light. Then the light from this star cannot overcome gravitational pull and escape into space, and will not reach a remote observer, so that we would be unable to see the star even though it does emit light!

If the mass of a celestial body is increased by adding to it matter of the same mean density, the escape velocity increases in proportion to the increase in radius (or diameter).

Now the conclusion drawn by Laplace is clear: for gravitation not to let the light out, a star of the Earth's density must have a diameter 250 times that of the Sun, that is, 27 000 times that of the Earth. Indeed, the escape velocity on the surface of such a star is also greater by a factor of 27 000 than on the surface of the Earth, nearly equal to the velocity of light. Hence, the star becomes invisible.

This was a spectacular prediction of one of the properties of the black hole: to confine light, and thus be invisible. To be fair, we should remark that Laplace was not the only one and indeed not the first to come up with this prediction. It was recently established that a similar proposition was made in 1783 by the British priest and geologist John Michell, one of the founders of scientific seismology. His arguments were very similar to those of Laplace.

A half-joking, half-serious (and sometimes serious) debate took place some time ago between the French and the British about who was to be honored as the discoverer of the possibility of existence of invisible stars: Laplace of France or Michell of Britain? Laplace's paper with the proof of the possibility of existence of black holes was cited in 1973 by the English physics theorists Stephen Hawking and George Ellis in a book devoted to special mathematical aspects of the structure of space and time; Michell's work was not known to specialists at the time. In the autumn of 1984, the English astrophysicist Martin Rees said at a conference in Toulouse that although what he had to say might not sound too nice in France, the truth was that the first man to predict invisible stars had been the Briton John Michell; Rees displayed a photograph of the title page of Michell's paper of 1784. This remark was greeted by the participants with applause and smiles.

Doesn't it resemble the debate of the French and the English about who predicted the coordinates of the planet Neptune on the basis of perturbations in the motion of Uranus, the Frenchman Le Verrier or the Englishman Adams? It is well known that they both calculated independently and correctly the position of the new planet. Le Verrier was luckier that time. Many a discovery has shared this fate. They are frequently made simultaneously and independently by several people. As a rule, a discovery is accredited to the one who demonstrated more profound comprehension of the problem, but sometimes it happens by a mere whim of fate.

Actually, the prediction made by Michell and Laplace was not yet the true prediction of black holes. Why not?

The point is that the science of the time did not yet know that nothing in nature is allowed to move at a velocity greater than that of light. Not in a vacuum, anyway. This was proved by Albert Einstein in special relativity theory but only in this century. For Michell and Laplace, therefore, the star that they considered was only black (nonemitting) since they were not aware that such a star loses any chance of 'communicating' with the outside world in any way, of 'informing' the world of any events occurring in the star. In other words, they could not know that it is not only 'black' but also a 'hole' into which one could fall but out of which there was no escape. Now we realize that if light cannot escape from some region of space, then nothing at all can emerge from it; we refer to this region as a *black hole*.

Another reason that detracts from the rigor of Michell's and Le Verrier's arguments is that they considered gravitational forces of enormous strength in which a falling body is accelerated to the velocity of light and the emitted light is confined, while at the same time applying Newton's law of gravitation.

Einstein was able to show that Newton's theory of gravitation is not valid for such fields, and developed a new theory, valid for superstrong and for rapidly varying fields (for which Newton's theory fails completely!); this theory is known as *general relativity*. If we want to prove that black holes can exist and to study their properties, it is this theory that we have to use as the tool.

General relativity is an astonishingly beautiful theory. It is so profound and elegant that everyone mastering it cannot but feel aesthetic delight. The Soviet physicists Lev Landau and Evgeny Lifshitz wrote in their textbook *The Classical Theory of Fields* that it is 'the most beautiful of all existing physical theories.' The German physicist Max Born said once that he enjoyed general relativity as he would an object of art. And the Soviet physicist Vitaly Ginzburg wrote that this theory makes him experience 'something akin to what one feels when contemplating the masterpieces of painting, sculpture, or architecture.'

No doubt, the numerous attempts at the popular exposition of Einstein's general relativity have created a general impression of its beauty. But let us be frank: the impression is as remote from the fascination imparted by the comprehension of the theory as a look at a reproduction of Raphael's *Sistine Madonna* is remote from the emotional experience you go through when contemplating the original created by the genius of the painter.

Nevertheless, if enjoying the original is impossible, one can (and ought to!) get acquainted with the available, preferably high-quality, reproductions (their merits do vary greatly).

A brief outline of some corollaries of Einstein's theory of general relativity is needed to make understandable the unbelievable properties of black holes.

Gravitational radius

What are the differences between Einstein's and Newton's theories of gravitation? Let us begin with the simplest case. Assume that we are on the surface of a spherical nonrotating planet and are to measure the attractive force exerted by this planet on some body, using a spring balance. We know that, according to Newton's law, this force is proportional to the product of the planet mass and the mass of the body, and inversely proportional to the squared radius of the planet. The radius can be found, for instance, by measuring the length of the equator and dividing it by 2π.

But what does Einstein's theory of gravitation have to say about this force? It predicts that the force is slightly greater than the value yielded by Newton's formula. We will later elaborate the meaning of 'slightly greater'.

Now imagine that we can gradually reduce the radius of the planet by compressing it while preserving its total mass. The gravity on the surface will increase (since the radius decreases). According to Newton, contraction by a factor of two increases the force fourfold. Einstein's theory says that the force will increase slightly faster: the smaller the planet radius, the greater the difference.

If the planet is compressed to such a degree that the gravity becomes superstrong, the difference between the value calculated in Newton's theory and the true value predicted by Einstein's theory starts to grow catastrophically. In the former theory, gravity tends to infinity as the body is compressed to a point (the radius is nearly zero). In the latter theory, the conclusion is very different: the force tends to infinity as the radius approaches the so-called gravitational radius. This value of the radius is determined by the mass of the celestial body: the smaller the mass, the smaller the gravitational radius. In fact, it is very small even for gigantic masses. Thus it equals only one centimeter for the Earth. Even the Sun's gravitational radius is only three kilometers. As a rule, the dimensions of celestial bodies are much greater than their gravitational radii. For example, the average radius of the Earth is 6400 km, and that of the Sun is 700 000 km. If the actual radii of bodies are much greater than their gravitational radii, the differences between forces calculated in Einstein's and in Newton's theories are extremely small. On the surface of the Earth, for example, this difference is one billionth of the value of the force.

The differences become appreciable only when the radius of the compressed body approaches the gravitational radius and the gravitational field becomes very strong; as mentioned above, the actual strength of the gravitational field becomes infinite when the radius of the body becomes equal to the gravitational radius.

Before discussing the consequences of this behavior, let us look at some other conclusions of Einstein's theory.

The essential point of the theory is that it connected into an inseparable whole the geometric properties of space and time and the gravitational forces. These connections are complex and diverse. We shall now single out two important features.

According to Einstein's theory, time in a strong gravitational field goes slower than time measured far from gravitating bodies (where gravitation is weak). Readers will certainly have heard that time can progress at a different pace. Nevertheless, this is something difficult to get used to. How can the pace of time vary? According to our intuitive feeling, time is the duration, something common to all processes. Time is like a river whose flow is unaffected by anything. Some processes may be faster or slower, and we can affect their rates by changing external conditions. For example, heating can accelerate chemical reactions, and freezing can slow down the functioning of an organism, but the motion of electrons in atoms will not change pace in response to these factors. It seems to us that all processes exist within the river of absolute time whose flow is unlikely to be affected by anything. Our notions allow the removal of all processes from this river, after which time will continue flowing as empty duration.

These notions reigned in Aristotle's and Newton's times, in fact, until Einstein. In Aristotle's *Physics* we find: 'Time flowing in two similar and simultaneous motions is the same time. If the two time intervals were not simultaneous, they would nevertheless be identical ... Hence, motions can be different and independent of one another. Time is absolutely the same in the former and latter cases.'

Newton, believing that he was speaking of a self-evident truth, wrote: 'Absolute, true, and mathematical time, of itself, and from its own nature, flows equally without relation to any thing external.'

A suspicion that the concept of absolute time is not so obvious was sometimes expressed even in ancient times. Thus Lucretius wrote in the poem *On the Nature of Things* in the first century BC:

E'en TIME, that measures all things, of itself
Exists not; ... for of TIME
From these disjoined, in motion, or at rest
Tranquil and still, what mortal can conceive.

It was Einstein who proved that absolute time is fiction. The flow of time depends on motion and, which is especially important for us here, on gravitational field. In the gravitational field all processes – absolutely all of them, regardless of their nature – slow down for an outside observer. This means that what is slowed down is that which is common to all processes: time.

The amount of this slowing down (time dilation) is usually very small. Thus time on the surface of the Earth ticks slower than in deep space by the same fraction of one part per billion that we had when calculating the force of gravity.

It is remarkable that this minute time dilation in the gravitational field of the Earth was directly measured. A similar effect in the gravitational field of stars was also measured, even though this effect is also extremely small. Time dilation becomes considerably greater in very strong gravitational fields and tends to infinity as the body radius becomes equal to the gravitational radius.

The second important conclusion of Einstein's theory states that strong gravitational fields change the geometric properties of space. Euclidean geometry, which we are so accustomed to, becomes invalid. This means, for example, that the sum of the angles in a triangle is not equal to two right angles, and the circumference of a circle is not equal to the distance from the center times 2π. The properties of ordinary geometric figures change as if they were drawn on a curved surface. It is for this reason that space in gravitational fields is said to become 'curved'. Obviously, this curving becomes appreciable only in strong gravitational fields, when the size of a body approaches its gravitational radius.

Undoubtedly, the image of curved space is as difficult to reconcile with our deeply rooted intuitive notions as that of different rates of time flow is.

Newton wrote about space in terms no less definite than those concerning time: 'Absolute space, in its own nature, without relation to anything external, remains always similar and immovable'. He thought of space as of an infinitely large 'scene' on which unfold 'events' that do not affect this 'scene' in any way.

Nikolai Lobachevsky, the discoverer of a non-Euclidean, 'curved' geometry, had already expressed the idea that not Euclidean but his, Lobachevsky's, geometry may manifest itself in certain physical situations. Einstein's calculations demonstrated that space is indeed 'curved' in a strong gravitational field.

This conclusion of the theory was also supported by direct measurements.

Why, then, is it so difficult to accept the conclusions of general relativity on space and time?

It is painful because the everyday experience of mankind, and even the experience of exact sciences, dealt for centuries with conditions under which changes in the properties of time and space were completely unnoticeable and thus were neglected. The body of our knowledge is founded on everyday experience. Hence, we are thoroughly used to the millennia-old dogma of absolutely unchanging space and time.

In our epoch, mankind became aware of conditions under which the effect of matter on the properties of space and time could not be ignored. We have to get accustomed to this peculiar situation despite the inertia typical of our thinking. New generations of people accept the predictions of general relativity much more easily than the generation of several decades ago when even the most enlightened minds had difficulty in mastering Einstein's theory. I will add another remark on the conclusions of relativity. Its author proved that not only the properties of space and time can change but that space and time merge into an inseparable whole: the four-dimensional 'spacetime'. It is this unified manifold that undergoes curving. Obviously, visual images are even more difficult to work out in this four-dimensional supergeometry; we will not spend time on it here.

Let us return to the gravitational field around a spherical mass.

We have to specify what is meant by the radius of a circle, for example, of a planetary equator, because the geometry in a strong gravitational field is non-Euclidean, curved. In conventional geometry there are two ways of determining the radius: first, as a distance from points on the circumference of the circle to the center, and second, as the length of the circumference divided by 2π. As a result of 'space curvature', these two quantities do not coincide in non-Euclidean geometry.

The second way of determining the radius of a gravitating body (not of the distance from the center to the circumference) has a number of advantages. To measure the radius, one need not approach the center of a gravitating mass. This is an important feature; for instance, it would be quite difficult to reach the center of the Earth when measuring its radius but not too difficult to measure the length of the equator.

In the case of the Earth, there is no need to measure the distance from the center directly: the gravitational field is not strong, Euclidean geometry holds for us with high accuracy, so that the length of the equator divided by 2π equals the distance from the center. This is not so, however, in superdense stars with strong gravitational fields: the difference between 'radii' determined by the two methods may be considerable. Furthermore, we will later see that in some cases the gravitational center is unreachable in principle. For this reason, we will always refer to the radius as the length of the circle divided by 2π.

The gravitational field around a spherical nonrotating body, discussed above, is known as the Schwarzschild field, after the scientist who solved the equations of general relativity for this case immediately after Einstein published his theory.

The German astronomer Karl Schwarzschild was one of the creators of modern theoretical astrophysics; he also made important contributions to practical astrophysics and to other branches of astronomy. Schwarzschild died when he was only 42 years old; at the session of the Prussian Academy of Sciences devoted to his memory, Einstein gave the following appreciation of Schwarzschild's contribution to science:

The most impressive characteristic of Schwarzschild's theoretical papers is a perfect command of mathematical tools of analysis and the ease with which he uncovered the core of an astronomical or physical problem. The combination of such profound mathematical knowledge and common sense with Schwarzschild's flexibility of reasoning is a very rare gift. It was this talent that allowed Schwarzschild to carry out important theoretical work in the fields whose mathematical obstacles scared away other theorists. It appears that the stimulus behind his inexhaustible creative effort was not so much the urge to perceive the hidden relationships of nature as the joy of an artist who discovers a subtle web of interconnections between mathematical concepts.

Schwarzschild obtained the solution to Einstein's equations for the gravitational field of a spherical body in December 1915. We have already mentioned that Einstein's theory, being based on completely novel, revolutionary concepts, is a very complicated one, but it is also extremely complicated 'technically'. Newton's formula of the law of gravitation is famous for its classical simplicity and brevity; in contrast, the new theory required that the gravitational field be found by solving a system of ten equations, each containing hundreds [*sic*] of terms. Moreover, these are not just algebraic equations but second-order partial differential equations.

Nowadays the entire gamut of electronic computers is employed to handle problems of this class. Of course, no such help was available in Schwarzschild's time, pen and paper being the only tools.

In fact, work in general relativity may require in certain cases – even today – a great many hours of painstaking mathematical manipulations 'by hand' (without computer assistance); they are often tedious and repetitive because of the staggering number of terms in the formulas. However, this crude labor is unavoidable. I often suggest to students (and sometimes to postgraduate students and young research fellows), fascinated by the brilliance of general relativity as taught by textbooks and wishing to work in it, that they start with calculating 'by hand' at least one relatively simple quantity arising in problems of this theory. Not every one of these young people strives to devote his or her life to general relativity after many days (sometimes months) of such calculations.

To defend this 'test for love', I should confess that I myself was subjected to it in this very manner. (Incidentally, legends show that ordinary love between two people also used to be tested by heroic deeds.) In my student days, my instructor in general relativity was a well-known specialist and exceptionally modest person, Professor Zelmanov. The problem he chose for my diploma assignment was connected with a fascinating property of the gravitational field: the possibility of 'cancelling' it anywhere, if we wish. 'Oh, no!' I hear the reader protest, 'Textbooks insist that gravitation cannot in principle be shut off by any screen, and that "cavorite" invented by H. G. Wells is pure fiction, forbidden by nature.'

All this is true; as long as we are at rest with respect to, say, the Earth, its gravitational pull cannot be eliminated. But the effect of this force can be completely cancelled if we start falling freely in this field. Free fall produces weightlessness. There is no weight inside a spaceship orbiting the Earth with engines shut down: astronauts and their equipment float in the cabin and feel no weight. We have watched this picture quite a few times on television screens. Note that no other field, for instance, electromagnetic, allows such simple 'cancellation'.

This property of gravitation is related to a very complex problem of the theory: the problem of the energy of the gravitational field. Some physicists are of the opinion that this problem has not yet been solved. The formulas of the theory make it possible to calculate for any mass the total energy of its gravitational field in the whole of space. However, one cannot specify where this energy resides, or how much energy is located at a specific point of space. In physicists' jargon, the concept of gravitational energy density at a point of space cannot be introduced.

The task of my diploma assignment was to prove by direct calculations that the mathematical expressions known at the time for gravitational energy density gave meaningless results even for observers not in free fall, say, for observers at rest on the Earth who definitely sense the force exerted on them by the planet. The mathematical expressions I was to operate with were even more cumbersome than the equations of the gravitational field discussed above. I even asked Professor Zelmanov to give me an assistant who would carry out the same calculations in parallel: I was afraid of overlooking a mistake. However, Professor Zelmanov knew (and I did not) that the ultimate goal was not only to obtain results of concrete calculations but also to bring up a young researcher. He said in his usual gentle manner, but quite definitely, that I would have to do the job single-handedly.

When everything was over, I realized that the routine work had taken several hundred hours. Almost all calculations had to be carried out twice, sometimes more. By the day the result of the diploma assignment was to be presented, the pace of the work grew rapidly, just like the velocity of free fall in the gravitational field. In all fairness, I have to say that the essential point of the work was not reduced to direct calculations. As it proceeded, I had to think of and solve problems of principle.

This was my first publication on general relativity.

Let us return to Schwarzschild's paper. Using elegant mathematical analysis, he had solved the problem for a spherical body and sent the solution to Einstein for presentation to the Berlin Academy. The solution fascinated Einstein because by that time he himself had been able to obtain only an approximate solution valid for weak gravitational fields. In contrast, Schwarzschild's solution was exact, that is, valid for arbitrarily strong gravitational fields around a spherically symmetric mass; this was a very important result. In fact, neither Einstein nor Schwarzschild himself knew at that time that this solution contained something much greater. It was later found that it contained the description of a black hole.

We shall resume the discussion of escape velocity. What velocity is to be imparted, according to Einstein's equations, to a rocket starting from the surface of a planet in order that it break away from the gravitational forces and escape into space?

The answer proved to be very simple: the formula given by Newton's theory remains valid in Einstein's case. Therefore Laplace's conclusion on the impossibility of light escaping from a compact gravitating mass was confirmed by Einstein's theory of gravitation which states that the escape velocity must become equal to the velocity of light right at the gravitational radius.

A sphere whose radius equals the gravitational radius is known as the Schwarzschild sphere.

Prediction

According to Einstein's theory, therefore, light is unable to leave the surface of a body and reach a distant observer once the radius of this body decreases to the gravitational radius; the body becomes invisible. However, the reader will of course have noticed that this extremely unusual property is certainly not the only 'miracle' that has to happen to a body whose size decreases to the gravitational radius. As we explained in the preceding section, the gravity on the surface of a star that reaches the gravitational radius must become infinitely large, and so must the acceleration of free fall. What will be the consequences of this situation?

To answer this question, recall first why ordinary stars and planets do not get compressed to a central point by the gravitational force but form equilibrium bodies.

The centerward compression is balanced out by the forces of internal pressure of matter. In the case of stars, this is the pressure of very hot gas that tends to expand the star. In Earth-type planets, these are the forces of tension, elasticity, and pressure that also counteract the compression. The equilibrium of a celestial body is maintained by this very equality of the forces of gravitation and forces counteracting it.

These latter forces depend on the state of matter, that is, on its pressure and temperature. Compression increases them. If, however, the matter is compressed to a finite (but not infinitely high) density, pressure and temperature remain finite as well. The gravitational force behaves differently. As the body size approaches the gravitational radius, the gravity on the surface tends, as we already know, to infinity. Now it cannot be balanced out by the finite counteracting pressure and the body must irresistibly contract towards the center.

An extremely important conclusion of Einstein's theory thus states: a spherical body whose radius reaches the gravitational value or smaller than this value cannot be at rest but has to contract centerward. 'But wait,' asks the reader, 'if gravity is infinitely high at the gravitational radius, what will it be when the body size drops to below the gravitational radius?'

The answer is fairly obvious. So far we have discussed the gravitational force on the surface of a static body that is not contracting at a given moment. But this force depends upon the state of motion. We have already said that free fall results in weightlessness: a freely falling body does not undergo gravitational force at all.

Hence, no gravitational pull is exerted on the surface of a freely contracting body (both within and without the Schwarzschild sphere). The matter forced to fall by gravitation cannot stop at the Schwarzschild sphere (otherwise it would be subject to infinitely high gravity). Stopping inside the Schwarzschild sphere is all the more prohibited. Any particle, say a rocket with no matter how powerful an engine, must fall on the center irresistibly once it has fallen to a point less than the gravitational radius from the gravitational center.

We have thus answered the question about the consequences of the infinite growth in the gravitational force as the body approaches the Schwarzschild sphere: it produces a catastrophic, unstoppable compression. Physicists refer to this phenomenon as the *relativistic collapse*.

It is thus sufficient to compress a body to the size of its gravitational radius for the further compression to be self-sustained. This process produces an object that is now known as a *black hole*.

The process of relativistic gravitational collapse described above was first rigorously calculated by the American physicists Robert Oppenheimer and Hartland Snyder in 1939. Their paper is a classic of succinct and lucid presentation. It gives a complete and stringent description of the phenomenon while occupying only a few pages.

Oppenheimer was well known far beyond the physics circles. He took part in the development of the American atomic bomb and headed the famous Los Alamos Scientific Laboratory in 1943–5. Later he realized the danger implied by the development of the hydrogen bomb and by the armaments race and spoke up for the utilization of atomic energy for peaceful purposes only; in 1953 he was stripped of all his governmental positions as a politically unreliable American.

The Oppenheimer–Snyder paper must be regarded as the rigorous prediction of the possibility of the generation of black holes. As for the term 'black hole', it was coined much later, at the end of the 1960s. Its inventor was the American physicist John Wheeler. In the USSR, for example, they were known for some time as 'collapsars'; this word was then rejected as non-euphonious in English. To be frank, the term 'black hole' led to dubiousness too, despite its precise and clear image.

In 1988 at an international conference in Leningrad, Professor Remo Ruffini of Rome University, who obtained many important results in black hole physics, and I recalled the initial stages of the stormy growth of this science. In 1972 several specialists, myself included, lectured on black hole theory at the International School in the vicinity of Les Houches in the French Alps. After the School ended, the lecturers got together to discuss details of publishing the lectures in a single volume. We had to choose the title of the volume. All agreed that the book was to be called *Black Holes*. Unexpectedly, the technical secretary of the school – a nice and pretty young woman from France – who was to prepare the texts for publication, blushed and said that this title would create serious difficulties. The point was that the book had to have on the title page the title both in English and in French (indeed, the entire text was in English and the school was organized in French). The secretary explained that *Black Holes* would look extremely odd in French. (Of course, we all communicated in English which long ago became the international scientific tongue and hardly any of the lecturers were fluent in French.) The secretary was absolutely adamant, stating that no publishing firm of high repute would print a science book with this title in French. We had to compromise. When published, the book had the English title *Black Holes* and the French title *Etoiles Noires* (*Black Stars*) which – you would agree – is something quite different. In fact, we will see later that this object is not only black, not letting any light out, but is precisely a hole in space and time!

We will conclude this chapter with the following remark.

In principle, a black hole could be produced artificially. We need to compress any mass to the size of its gravitational radius, after which it will contract of itself, undergoing gravitational collapse.

Actually, this endeavor would meet with enormous technical difficulties. The smaller the mass we want to transform into a black hole, the tinier the size to which it has to be compressed, since the gravitational radius is directly proportional to the mass. Thus we know that the gravitational radius of the Earth is roughly one centimeter; a mountain of, say, one billion tonnes would be converted into a black hole if compressed to the size of an atomic nucleus!

Subsequent chapters will show that large masses may spontaneously transform into black holes in the course of the natural evolution of the Universe. Before coming to that, however, we will continue outlining the fantastic peculiarities of black holes.

2: Around a black hole

A hole in time

We have mentioned that the theory of gravitation predicts that the closer the clock is to the gravitational radius the slower the flow of time will be. This means that no matter what processes proceed in a strong gravitational field, an observer located far from a black hole will see that their pace has slowed down.

Thus he will find that oscillations in atoms that emit light are slowed down so that photons originating at these atoms reach the observer 'reddened', at a reduced frequency. This phenomenon is called the gravitational red shift (it served as the basis for one of the tests of the validity of Einstein's theory). What is important for us for the moment is that the closer the emission region lies to the boundary of a black hole (to the Schwarzschild sphere) the greater the time dilation and reddening of light will be. Here time gets slower and then 'stands still' for a distant observer. If the observer follows a stone falling onto a black hole, he will find that close to the Schwarzschild sphere the stone starts to 'decelerate' and gets to the black hole boundary only after an infinitely long time.

A distant observer will see a similar picture in the very process of generation of a black hole, when the stellar matter is pulled by gravitation towards the center. For this observer, the surface of the star takes an infinitely long time to reach the Schwarzschild sphere as if stalling at the gravitational radius. This behavior suggested the now outdated term 'frozen star' for black hole.

This stalling does not mean that the observer is doomed to contemplate eternally the star surface anchored at the gravitational radius. Let us recall time dilation, and reddening of light emitted from regions of strong gravitational field. As the stellar surface approaches the gravitational radius, the observer receives gradually redder and redder light, even though the photons created on the surface itself are unchanged. Less energetic ('reddened') photons arrive at the observer at progressively greater intervals. The intensity of light decreases.

The red shift due to time dilation caused by a strong gravitational field is aggravated by reddening due to the Doppler effect. Indeed, the surface of the contracting star is constantly receding from the observer, and we know that the light emitted by a source moving away seems to be reddened.

The joint action of the Doppler effect and time dilation in the strong gravitational field finally makes the star invisible: as the stellar surface approaches the Schwarzschild sphere, the light reaching the distant observer becomes progressively more red and less intensive. Its brightness tends to zero and it becomes undetectable by any telescope. This fading off appears as practically instantaneous to the distant observer. Thus a star of one solar mass, having contracted to twice the gravitational radius, burns out for the distant observer in one hundred-thousandth of a second.

Radar would also be useless for detecting the surface of the star frozen at the gravitational radius. Radio waves will travel infinitely far to the gravitational radius and will never return to the observer who sent them. The star completely 'disappears' for distant observers, and only its gravitational field remains. An outside observer will never see what happens to the star after it has contracted to a size below the gravitational radius.

'Now hold on', says the reader. 'What's the use speaking about size below the gravitational radius if the process of contraction to the gravitational radius lasts for ever? You have just explained that the star stops contracting at dimensions equal to the gravitational radius. When is it that the star diminishes to less than the gravitational radius? After an infinitely long time?'

This point reveals one of the most puzzling and important truths discovered by general relativity: the relativity of temporal intervals, their dependence on the state of motion of the observer. Recall that, for different observers, the same process has different durations in special relativity: to a terrestrial observer, a clock in a rapidly moving rocket ticks slower than his own. This effect was confirmed by a direct physical experiment. If a clock falls on a black hole, the relative nature of the duration of the process manifests itself in a most fascinating way. Let us have a detailed look at it.

Imagine a string of observers spaced along a line that originates at the center of a black hole; the observers are at rest with respect to the black hole. For example, they may sit in rockets whose engines are running and keep the observers from falling into the hole. Now imagine another observer who is falling freely onto the black hole, in a rocket with the engine shut down. As the rocket falls, the observer moves past the observers at rest at a progressively greater velocity. If this observer falls from a great distance, his velocity equals the escape velocity. As the falling body approaches the gravitational radius, its velocity tends to that of light. Clearly, as velocity increases, the rate of time flow on a freely falling rocket slows down. The decrease is so large that for any observer on a fixed rocket the time necessary for the falling observer to reach the Schwarzschild sphere is infinitely large, while the time by the clock of the falling observer is finite. The infinitely long time of an observer on a fixed rocket is thus equal to a finite time interval of another observer (in the falling rocket); this interval is quite short: we saw that for one solar mass it equals only one hundred-thousandth of a second. Could there be a more illustrative example of the relative nature of duration?

We have found that according to the clock placed on the contracting star, the star contracts to its gravitational radius within a finite time interval and continues to contract still further. However, we remember that a distant outside observer can never see these last stages of evolution. And what about an observer on the contracting star after it has disappeared into the Schwarzschild sphere? What is the fate of the star?

Let us put off these questions for the time being and turn to the external field of the black hole in order to see how bodies move and light rays propagate in this superstrong field.

Celestial mechanics of black holes

According to Newton's theory of gravitation, a body in the gravitational field of a star moves along either an open curve (a hyperbola or a parabola) or a closed curve (an ellipse), depending on whether the initial velocity of motion is large or small. At a large distance from a black hole, its gravitational field is weak and all phenomena are described to a high degree of accuracy by Newton's theory; this means that in this field the laws of Newtonian celestial mechanics hold. As we approach a black hole, however, these laws are more and more violated.

Let us look at some important peculiarities of motion of bodies in the gravitational field of black holes.

By Newton's theory, if the velocity of a body is less than the escape velocity, it moves along an ellipse around the central body, that is, around the gravitation center. The ellipse has a point closest to this center (the periastron) and a point farthest away from it (the apoastron). By Einstein's theory, if the body moves at a velocity below the escape velocity, its trajectory also has a periastron and an apoastron but it is not an ellipse any more; the body follows an open orbit, alternately getting closer and then farther from the black hole. The entire trajectory lies in one plane but it may have a very intricate shape, as illustrated in Fig. 1. If the trajectory is sufficiently far from the black hole, it is an ellipse that slowly rotates in space. This slow rotation of the elliptic orbit of Mercury by 43 seconds of arc per century gave the first confirmation of the correctness of Einstein's theory of gravitation.

It is very interesting to discuss the simplest periodic circular motion of a body in the field of a black hole. By Newton's theory, circular motion is allowed at any distance from the gravitation center. Einstein's theory shows that this is not so. The closer the body to the gravitation center, the higher the velocity of motion on the circular orbit. This velocity reaches the velocity of light on the circle of one and a half times the gravitational radius. Circular motion on an orbit closer to the black hole is impossible since the body would have to move at a velocity above the velocity of light.

In fact, circular motion around a black hole in realistic situations is found to be impossible at greater distances as well, beginning with three gravitational radii, where the velocity of motion is only half the velocity of light. Why?

The thing is that circular motion at distances below three gravitational radii is unstable. The smallest perturbation, an arbitrarily small push, would force the body to leave the orbit and either fall into the black hole or fly away into space (nothing of the sort is predicted by Newtonian celestial mechanics). But the most interesting and unusual prediction concerning celestial mechanics is the possibility of gravitational capture by a black hole of bodies arriving from space.

You remember that in Newtonian mechanics any body flying in from space to a gravitating body traces around it a parabola or hyperbola and (unless the body 'strikes' the surface of the gravitating mass) flies away into space: no gravitational capture is possible. The situation is different in the field of a black hole. Obviously, if the body travels very far from the black hole (at a distance of tens of gravitational radii or more) where the gravitational field is weak and the laws of Newtonian mechanics hold, it follows almost exactly a parabolic or hyperbolic orbit. But if it moves sufficiently close to the black hole, its trajectory is very different from a parabola or hyperbola. If the velocity of the body far from the black hole is much less than the velocity of light and the orbit passes close to the circle of twice the gravitational radius, the body will make several revolutions

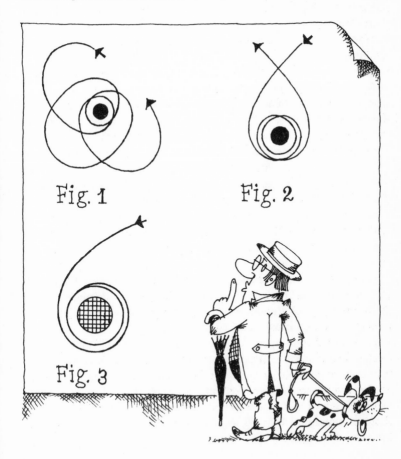

Fig. 1   Fig. 2

Fig. 3

around the black hole before flying away. This situation is shown in Fig. 2.

Finally, if the body comes close to the circle of twice the gravitational radius, its orbit will 'wind up' on this circle (like a thread on a spool); the body will be gravitationally captured by the black hole and will never leave it (Fig. 3). If the body comes even closer to the black hole, it will fall into the hole and thus also become gravitationally captured.

One more remark is needed before we pass on to other physical phenomena in the gravitational field of a black hole. It concerns the escape velocity. We have already mentioned that this velocity obeys the formula of Newton's theory, so that a body having this or higher velocity escapes from the pull of the black hole for ever. However, one qualification is needed.

Obviously, if a body moves towards the black hole exactly along the radius, it will hit the black hole whatever its initial velocity, and will never re-emerge.

Moreover, we know by now that if a body does not move along a radial ray but its orbit passes sufficiently close to the black hole, its fate is gravitational capture. Hence, for a body to break away from the vicinity of a black hole, it is not enough to have a velocity greater than the escape velocity; it is also necessary that this velocity be at an angle to the direction towards the black hole greater than a certain critical value. If the angle is less than critical, the body will be gravitationally captured, and if it is greater (and the velocity exceeds the escape value), the body flies away into space. The value of this critical angle depends on the distance from the black hole. The farther it is, the smaller the critical angle. If the body is at a distance of several gravitational radii, a very accurate 'aim' must be taken at the black hole for capture to occur.

Finally, a few words about another important process accompanying the motion in the field of a black hole. This phenomenon is the radiation of gravitational waves. Einstein's theory of gravitation predicts that they must exist.

What are these waves with such an exotic name? They resemble electromagnetic waves, that is, a rapidly oscillating electromagnetic field that 'broke loose' from its source and propagates through space at the maximum allowed velocity, that is, the velocity of light. Likewise, gravitational waves are the oscillating gravitational field that has 'separated' from its source and propagates through space at the velocity of light.

It is common knowledge that in order to detect an electromagnetic wave it is sufficient, in principle, to take an electrically charged ball and follow its motion; once an electromagnetic wave starts falling on the ball, it begins vibrating. In fact, one ball is not enough to detect gravitational waves. At least two balls a certain distance apart are needed (charging these balls would certainly be unnecessary). When a gravitational wave falls on them, the balls alternately move closer and then farther apart. Gravitational waves can be detected by measuring the separation between the balls. The reader might ask why a single ball is not enough.

The reason is as follows. If no other forces act on a ball, it floats in the field of the gravitational wave in a state of weightlessness. The ball senses no gravitational forces, so that a travelling gravitational

wave cannot be detected. This situation is identical to that we find in
a spaceship on its orbit. When astronauts are weightless, they
cannot detect, nor measure, the gravitational field. Two balls separated by a certain distance are located in slightly different gravitational fields and undergo slightly unequal accelerations, which leads to
relative motion that can be measured.

A charged ball is not really necessary for the detection of electromagnetic waves since various types of electromagnetic antennas are
available. A number of gravitational antennas have also been
designed to detect gravitational waves.

Actually, the detection seems to be relatively simple only in theory.
In reality, gravitational waves in any conditions that we are used to
are exceptionally weak: they are emitted in accelerated motions of
massive bodies. The radiation of gravitational waves must be negligibly small even in the motion of celestial bodies. Thus the motion of
planets in the Solar System generates gravitational energy whose
power equals that of a mere hundred electric light bulbs. Even
though this power may seem appreciable by our terrestrial standards, it is absolutely negligible if compared, say, with the power of
the light emitted by our Sun which is greater by a factor of one
hundred-thousand billion billion (a 'one' with twenty-three noughts,
that is 10^{23}). Attempts to devise laboratory emitters of gravitational
waves are doomed to failure.

For instance, the emitter of gravitational waves can be constructed
as a rapidly rotating rod. If we take a steel beam 20 m long with a
mass of 500 tonnes and make it revolve at a rotation rate just below
the point of rupture by centrifugal forces (rotation frequency of about
30 hertz), this device will emit a mere one-ten-thousandth of a
billionth of a billionth fraction of one erg per second.

These examples show the obstacles on the way to detecting
gravitational waves. So far, direct experiments in terrestrial laboratories have failed to observe these waves, although dozens of
gravitational antennas for signals of cosmic origin have been built
and are being built in a number of laboratories around the world.
The pioneer of this work was the American experimenter Weber at
the end of the 1950s and beginning of the 1960s. In the USSR, the
development of gravitational antennas is pursued most actively by
Braginsky's group at Moscow University.

Even though terrestrial antennas have so far failed to detect
gravitational waves, some astronomical observations directly indi-

cate that the motion of celestial bodies does lead to the emission of these waves. What are these observations?

We already know that as planets or, for instance, stars in binary stellar systems move along their orbits, they emit gravitational waves that carry away energy. As a rule, these losses are very, very small. However, the greater the mass of celestial bodies and the smaller the separation between them, the more intensive the emission of gravitational waves. Energy losses in a binary stellar system result in the gradual approach of the two stars and in a reduction of the period of revolution around the center of mass. Of course, this is a very slow process; nevertheless, special methods of observation have made it possible to discover in one case a reduction of period in exact agreement with the prediction of Einstein's theory. Here we will not describe in detail these astronomical observations since they would lead us too far astray.

Let us return to the motion of a body on a circular orbit around a black hole. It is accompanied by the emission of gravitational waves and the gradual reduction of the radius of the orbit. This process will last until the radius diminishes to the level of three gravitational radii. We already know that the motion at smaller distances is unstable. Hence, having reached the critical orbit and having made several more turns, the body emits a certain amount of energy and 'plunges down' from this distance into the black hole.

How much energy will be emitted by the body in the form of gravitational waves in the course of motion around the black hole on a circle of slowly decreasing radius? We saw that the intensity of this emission is extremely low; however, this process lasts for a very long time! The total amount of emitted energy will be large. We can estimate this energy using the following comparison. We know that a certain fraction of mass in nuclear transformations, for example, in the transformation of hydrogen into helium or heavier elements, is converted into energy. In all types of reaction this fraction is at most one per cent. However, in the case of the emission of gravitational waves in the motion around a black hole the percentage is six times as high!

We find that in principle black holes could be used as an energy source even in this simplest fashion. Of course, this device is useless for any practical purpose. The thing is that the interaction of gravitational waves with matter is unimaginably weak. The energy released in the form of gravitational waves would be extremely difficult to collect and put to practical use: gravitational waves would disperse throughout cosmic space. We will see later that there exist other ways of utilizing the tremendously high gravitational energy of black holes.

Black holes and light

We already know that the gravitational field acts on light. It changes the frequency of photons and bends the trajectories of light rays: the closer to a black hole, the more the trajectory is bent. Fig. 4 shows the paths of light rays emitted from points at different distances from a black hole (at right angles to the radius). Note the critical circle of one and a half times the gravitational radius. (It has already been mentioned in the preceding section.) A photon confined to this circle by the powerful pull of the black hole is free to propagate along the circle. This motion is, however, unstable. In response to any minute perturbation the photon either falls into the black hole or flies away into space.

The critical circle for photons implies that the light passing sufficiently close to the black hole is gravitationally captured by it (this is shown in Fig. 5). A ray coming very close to the circle of 1.5 gravitational radii winds up for an infinitely long time on this circle, and a ray coming even closer ends in the black hole.

Propagation of light in the vicinity of a black hole changes the frequency of light waves: the closer to the black hole, the greater the increase in oscillation frequency. When light travels away from the black hole, its frequency decreases. These changes are appreciable only in the neighborhood of the Schwarzschild sphere, being very small at considerable distances from the black hole.

'Black holes have no hair'

So far we have discussed only black holes that are formed in the compression of spherical bodies and thus have a spherically symmetric gravitational field. What would a black hole be like if not a spherical but, say, a flattened body were contracting? Leaving the role of rotation to the next section, we will be concerned here only with nonrotating bodies.

Assume that, prior to compression, a body had a nonspherical gravitational field. Does it mean that collapse will result in a flattened black hole with a flattened gravitational field? The answer to this question was unknown for a long time, and the problem was solved fairly recently. It was shown that there cannot exist flattened or any other type of nonsymmetrical black hole. The point is that when the size of the body approaches the gravitational radius in the course of contraction, intensive emission of gravitational waves occurs. All deviations of the gravitational field from the strictly spherical shape are shown to 'shrink' and be 'radiated away' in the form of gravitational waves.

In the first moments after birth, the black hole does have a distorted, flattened shape. But this shape cannot survive for long. Like the wall of a soap bubble that rapidly returns to the spherical form when we stretch it and let go, the boundary of a 'distorted' black hole very rapidly regains its smooth spherical shape. All 'redundant' details are radiated away as gravitational waves. The result is a perfectly symmetric black hole generating a perfectly

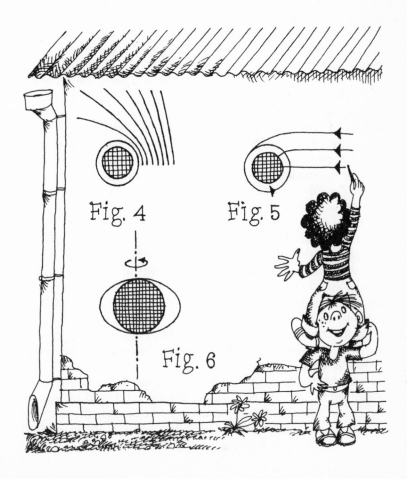

symmetric Schwarzschild gravitational field characterized by a single quantity, that is, the mass of the gravitating center.

Black holes thus can be bigger (more massive) or smaller but are quite identical in all other respects. The question that immediately arises then is what happens if the contracting body has some electric charge, that is, if it is surrounded (in addition to the gravitational field) by an electric, magnetic, or some other field? Will the black hole born of this body also possess these fields?

An analysis of this problem led to an extremely interesting conclusion. It was found that all types of physical field are radiated away or buried in the black hole during the relativistic collapse. The only exception is the electric charge. The field of this charge, just as the spherical gravitational field, does not change at all and continues to surround the black hole.

We have thus seen that the relativistic collapse of any nonrotating body, no matter what its shape, when surrounded by an electric, magnetic, or other field results in the birth of a black hole whose properties are completely characterized by only two parameters: the mass determining the strength of the external gravitational field, and the electric charge determining the electric field. All other distinctive features of matter that formed the black hole seem to disappear. No measurements or experiments with a black hole can reveal whether it was made up of matter or antimatter, whether or not it originally had a magnetic field, and so forth. This property of 'forgetting' all distinctive attributes is also rooted in the fact that no signals escape from the black hole to external space.

Putting aside the phenomenon of electric charge which is relatively insignificant for celestial bodies, we are left with mass as the only characteristic that determines the properties of a black hole. All black holes of identical mass are exact copies of one another. This facelessness of black holes led the American physics theorist John Wheeler (already familiar to the reader) to the aphorism 'Black holes have no hair.'

To establish this facelessness was a difficult task for astrophysics. It also has its prehistory and its history, just as the theoretical prediction of black holes.

I will mention only two papers presenting the results of work carried out in the mid 1960s. I was a witness to the work on the first of them and participated in the work on the second. The first was written by the Soviet physicist Vitaly Ginzburg. He wanted to calculate the magnetic field of a star if it is compressed to ever smaller radii. It was found that if the star is squeezed to almost the gravitational radius and the process is stopped, the magnetic field very close to the surface of the star is unusually enhanced. Subsequent compression to exactly the gravitational radius would increase the strength of the magnetic field at the surface to infinity. But this is absurd! I clearly remember the enthusiasm and agitation with which Professor Ginzburg reported this result at the P. K. Shternberg Astronomical Institute, and the happy enthusiasm of the specialists discussing the conclusions. The importance of this result was tremendous. Indeed, if the assumption of a nonzero magnetic field around a black hole produces an absurdity (this was indeed so!), then a black hole cannot have any magnetic field at all! The magnetic field must be entirely radiated away or buried inside the black hole!

This was a very unexpected conclusion, and it took specialists quite some time to get used to it.

The second paper concerned the possibility of forming a flattened black hole. At that time I had just graduated from the University and had begun working in a group assembled by Professor Yakov Zel'dovich. I had already been excited for several years by the problem of a flattened black hole, and discussed it both with my Professor and with my peer, Andrei Doroshkevich, who had also begun to work in the newly created group. The three of us began a comprehensive attack on the problem. Each had his own way of working, fortunately complementary, and we managed to cover a good deal of ground.

We soon found that if the newly formed black hole were flattened as a turnip or elongated as a cucumber, this flattening or elongation had to be infinitely large! In the case of flattening, for example, this means that the length of the equator of the black hole must be infinite. This would be nonsense. We had to conclude that neither flattened nor elongated black holes can form. Any deviation from spherical shape must be radiated away as gravitational waves and vanish.

The two papers I have described have an important common feature: they predict that the formation of black holes must involve the radiation of gravitational waves. This conclusion may seem to be unjustifiably daring. Indeed, we did not calculate the radiation process as such: we simply couldn't do it at the time because the required mathematical techniques had not yet been developed. But the conclusion on the result of the radiation process was absolutely reliable because a different one would yield an absurd outcome. This is an interesting example of how one can draw correct conclusions on the consequences of a phenomenon that is too complicated to be calculated. Only six years later the American theorist Richard Price, and later many others, succeeded in calculating the process of radiation of the fields. As could be expected, the results confirmed the correctness of our conclusions.

Price's and others' calculations revealed the following unusual fact: all fields that can be radiated away are indeed radiated in the course of relativistic collapse. Only two types of field are never shed: the spherical gravitational field and the spherical field of electric charge (if the black hole is charged). These are the fields retained by the ultimate black hole.

The next section will describe one more exception.

Physicists are familiar with 'Chisholm's laws'. They reflect in a jocular form the no-joke difficulties encountered by experimenters in their work. The first of Chisholm's laws reads: 'All things that can break down, do break down.' By analogy, Richard Price formulated his conclusion in the following form: 'All things that can be radiated get completely radiated away ("What is permitted is compulsory").'

We specially qualified at the beginning of this section that our subject was exclusively a black hole born of a nonrotating body. This qualification was not accidental. Namely, a rotating collapsing body generates a rotating black hole. We will see in the next section that rotation causes changes in the gravitational field and thus consti- tutes the third (and last) parameter (on top of mass and electric charge) that characterizes the black hole.

Gravitational vortex around a black hole

By Newton's theory, a gravitational field is in no way dependent on the motion of matter. Thus the gravitational fields of a revolving sphere and a sphere at rest are perfectly identical provided their masses are equal. By Einstein's theory, this is not so: the gravitation- al fields of the two spheres differ slightly. What is this difference?

The most illustrative (if somewhat oversimplified) picture of this difference is the rotating body surrounded by an additional vortex gravitational field which entrains all bodies into circular motion. Another illustration would be to say that layers of space slowly revolve around the body, with the angular velocity of rotation being a function of distance: it is low far from the body and increases as we move closer to it. These effects are negligible for ordinary celestial bodies; the easiest way to detect them is by placing a gyroscope in the vicinity of the rotating body. If the body is at rest, the gyroscope points in a direction that does not change with respect to distant stars. (The use of gyroscopes for, say, orientation of space probes is well known.) Close to the rotating body, however, the gyroscope's axis slowly rotates. Thus a gyroscope close to the surface of rotating Earth turns by about one-tenth of one second of arc per year. Of course, this minute rate of rotation of a gyroscope cannot disrupt the orientation of spaceships. Moreover, so far this effect has not been observed experimentally.

The angular velocity of rotation of a gyroscope's axis may be very high at the surface of neutron stars mentioned in Chapter 1, only several times smaller than the rate of rotation of the neutron star itself. These stars may revolve at a rate of several tens of revolutions per second or more. A gyroscope in the vicinity of such a rapidly rotating star may revolve at a rate of many revolutions per second! What is going to happen to this vortex component of the gravitational field when the star undergoes relativistic collapse?

It has been shown that just as the spherical gravitational field remains unchanged, so this component remains invariant.

The vortex gravitational field of a star is completely determined by the quantity that physicists call the angular momentum of the body. For an ordinary star, this quantity is approximately equal to the product of the rotation velocity at the equator, the stellar mass, and the stellar radius.

The collapse of a rotating body thus generates a rotating black hole. What does it mean: the rotation of a black hole? It means that the black hole is surrounded by a vortex gravitational field left behind by the collapse; sometimes this field is called the gravitational vortex. The closer to the black hole, the stronger the vortex field.

What are the consequences of this?

First, rotation somewhat flattens the black hole at the poles, just like rotation flattens the Earth and stars. You remember that if rotation is absent, the black hole is exactly spherical. But this is not the main aspect. Without rotation, gravitational force becomes infinite on the Schwarzschild sphere. This sphere constitutes the boundary of the black hole or, as physicists say, the horizon that does not let anything out. Rotation changes this behavior. Gravitation becomes infinite outside the horizon, on the surface called the ergosphere. This bounding surface is drawn in Fig. 6. The faster its rotation, the farther it lies, but it cannot go too far away from it.

No force can keep a body at rest if it reaches the boundary of the ergosphere or crosses it. The vortex field entrains it into a motion with respect to the black hole. However, in contrast to bodies under the Schwarzschild sphere (in the case of no rotation) that are pulled irresistibly towards the center, all bodies under the ergosphere surface are entrained into rotational motion around the black hole. They are not necessarily approaching the center: they may come closer, or recede, or may cross the ergosphere, moving in and out of it.

The question that arises here is: how is a body under the ergosphere surface affected by gravitational force if this force has already become infinite on this surface?

I must repeat the argument used in the discussion of the force acting on a body on the Schwarzschild surface.

The gravitational force is infinite on the surface only for a body at rest; if the body accelerates, the force is different. If the body circles the black hole in the direction of rotation of the latter, the force is found to be finite both on the surface of and inside the ergosphere. As a result, a body can move inside the ergosphere on a circular path without falling on the center. This means that rotation drastically changes the static limit (i.e. the boundary of the region within which a body can stay at rest with respect to the black hole) in comparison with the Schwarzschild sphere.

The ergosphere surface is not, therefore, the black hole boundary because a body is allowed to escape from under it. Let us see what happens when the body moves closer to the black hole.

On the way in, we finally reach the boundary of the black hole: its horizon. Bodies on this surface and below it (any particles and light photons) move only into the black hole. Outward motion is forbidden, no information from within the horizon can reach outside observers.

This space between the horizon and the static limit is called the ergosphere. Here the gravitational force makes all bodies circle the black body.

If a gyroscope is slowly brought closer to the surface of the ergosphere, its angular velocity will increase, tending to infinity on the surface (for a gyroscope at rest).

What will a distant observer see if a body is falling on a rotating black hole from a large distance?

In its fall on a black hole, the body's trajectory first deviates in the direction of rotation, intersects the ergosphere, and gradually approaches the horizon. At whatever point on the horizon a falling body arrives, its angular velocity is always the same. This is a very important property of rotating black holes. Angular velocities of bodies inside the ergosphere can vary but once they reach the black hole surface the bodies have identical angular velocities, and rotate together with the black hole surface as if glued to the surface of a rotating solid.

For an outside observer, the light emitted by these bodies shifts more and more to the red end and becomes less intense; then fades out completely, and the bodies become invisible to distant observers: they cannot see what goes on below the horizon. If, however, the observer is in free fall on a rotating black hole, he reaches the horizon in a finite time, as in the case of a nonrotating black hole, and continues to fall into it. We now leave this observer alone and return to the external space, that is, to the neighborhood of the black hole.

The rotation of a black hole cannot be arbitrarily rapid. The point is that if the parent body rotates too fast, the black hole cannot form. Indeed, the contraction of a body that rotates sufficiently fast produces at the equator the centrifugal forces that stop the compression in the equatorial plane. The body can continue contracting only at the poles. Hence, it transforms into a 'pancake' with a radius much greater than the gravitational radius, so that a black hole is not born. The rotation of a black hole reaches the maximum possible rate when the velocity of points on the equator becomes equal to the velocity of light.

The laws of celestial mechanics also undergo changes at a rotating black hole. For instance, take the effect of gravitational capture of bodies by a black hole. If the hole is rotating, the easiest to capture are particles that move counter to the rotation of the hole in its vicinity, while particles moving in the opposite direction (in the direction of rotation) are the hardest to capture. A suitable image is that of the vortex component of the gravitational field around the black hole acting as a sling: it accelerates and throws off those particles that pass by the black hole in the same direction as the vortex of this field, but it decelerates and captures those particles that move counter to this vortex.

Here is one more example of changes in the laws of celestial mechanics. If a body revolves on a circular orbit around a black hole at the maximum possible rate of revolution, it emits seven times as much energy in the form of gravitational waves as it would in the motion around a nonrotating black hole.

3: A gravitational abyss as an energy source

Bottomless black holes

We have mentioned several times that the radiation of gravitational waves by a body circling a black hole is a method of generating energy. However, this is not a way of extracting energy from the black hole, but rather the energy of the orbiting body. Indeed, ultimately the body (and part of gravitational waves) will fall into the black hole and therefore increase its mass and, consequently, its energy.

The question that arises is: is it possible to invent a process that reduces the mass of the black hole and thus extracts energy from it?

At a first glance, this seems to be impossible since a black hole does not let anything out, and hence, no energy can be extracted from under the horizon. This assumption is correct. But this reasoning leaves out the fact that part of the energy (and hence, part of the mass) of the rotating black hole, related precisely to rotation, resides 'outside' the black hole, in the vortex component of its field. It is found that this 'rotational' part of the energy can indeed be 'pumped out' of the black hole, thus reducing its mass. How can this be accomplished?

Imagine the following experiment. A rocket with its engines shut down gets into the ergosphere of a large rotating black hole. It moves around the black hole in the direction of its rotation. The pilot switches on the engines in the immediate vicinity of the black hole; exhaust gases emerge. The motion of the rocket can be changed, so that the exhausts fall into the black hole while the rocket is accelerated and ejected from the ergosphere at a tremendous velocity, as if launched by a 'sling', that is, a gravitational vortex. The velocity of the rocket will be much greater than the approach velocity, and much greater than the change in velocity due to the short burst of engines. What has happened?

You will remember that a rotational gravitational vortex exists around the black hole. The rocket engine has forced the rocket into a new orbit where it has been gripped by this vortex and hurled at a colossal velocity out of the ergosphere. The energy carried away by the rocket is supplied by the vortex, that is, it comes from the 'rotational' energy of the black hole. As a result, the rotation of the black hole slows down. Correspondingly, the total mass of the black hole also decreases (by the amount carried off by the rocket). This is the manner in which energy can be extracted from black holes.

This extremely unusual process was discovered by the English physics theorist Roger Penrose. We have already emphasized that the process draws only the 'rotational' energy that resides in the vortex field outside the black hole.

The area of the horizon (it is this quantity that characterizes the size of the black hole), is slightly increased by the process described because the gases from the rocket engine fall into the black hole, increasing its mass and hence, its size.

The rocket can carry away the maximum amount of 'rotational' energy from the black hole (for the same duration of the engine switch-on pulse) if the engines are turned on exactly at the horizon. In this case the horizon area remains unchanged (such processes are called reversible processes). Such manipulations with the engines at the horizon can be repeated an indefinite number of times, so that finally the black hole loses its 'rotational' energy but retains the size of its horizon.

A question about the possibility of a process that would reduce the horizon size has to be answered in the negative. It has been found that the area of the black hole horizon cannot be reduced by any process. If several black holes interact, the sum of the areas of their horizons cannot decrease.

This is a very important property, established by the famous English physicist Stephen Hawking. It implies, for example, that a black hole can never split into a pair of daughter black holes. If this did happen, the sum of horizon areas of the daughter holes would have to be smaller than the area of the parent black hole, but with the black hole energy conserved. As a result, no gravitational tide force or any other factor could 'tear' the black hole apart.

Merging of black holes is possible. For instance, two black holes colliding 'head on' coalesce into one. The resultant black hole has a horizon area that is greater than the sum of horizon areas of the colliding holes.

No process is thus allowed that reduces the size of a black hole.

Once born, a black hole is something like a bottomless precipice that cannot be filled or 'plugged' with anything: they are 'perpetual holes' in space and time that can only grow at the expense of matter falling into them. These are ever-growing gravitational abysses.

Actually, the situation, as we describe it somewhat later, is not so grim. First, the gigantic gravitational field of black holes can stimulate extremely stormy processes in real conditions; second, quantum processes (we have ignored them so far) introduce modifications into the picture outlined above. This subject is taken up later in the book.

Gravitational bomb

In our analysis of processes around a black hole and methods of extracting energy from it, we have found that this energy can be extracted either in the form of radiated gravitational waves or as the kinetic energy of bodies ejected from the ergosphere. It was discovered, however, that even more surprising and unpredictable approaches to exploiting black holes as generators of energy can be pointed out.

Imagine that a rotating black hole is irradiated with electromagnetic waves. What will the result be? At a first glance, nothing of interest will occur. The waves will be partly captured by the black hole and will vanish forever. Other waves passing by the black hole will have their trajectories bent, and will propagate away. The phenomenon in which the direction of propagation is changed is known as the scattering of waves. Scattered electromagnetic waves leaving the black hole have the same frequency that they had on the way to it. Of course, the frequency of the waves did change in their motion through the strong gravitational field at the black hole. It was increased on the way towards the black hole and their energy increased: this was the blue shift. On the way from the black hole, the waves undergo a red shift, and their frequency returns to the initial value far from the black hole.

On the whole, the picture is as follows. When a black hole is irradiated, electromagnetic waves partly sink into it but are partly scattered at a frequency that they had prior to scattering. Since waves are partly captured by the black hole, the intensity of the scattered waves is lower than the original intensity of the irradiating beam.

So far the picture is trivial. A situation is, however, possible in which the intensity of scattered electromagnetic waves exceeds that of the incident waves. This effect is possible if, first, the black body is rotating, since only rotational energy can be extracted. Secondly, it is necessary that the frequency of electromagnetic waves incident on the black hole is lower than the frequency of rotation of the hole. This is the situation in which the scattered electromagnetic waves have greater intensity than the incident beam. This process of amplification is known as superradiance. It was discovered by Professor Zel'dovich. Superradiance is in fact similar to the process discussed earlier of enhancement of the energy of a body ejected from the ergosphere and 'feeding' on the 'rotational' energy of the black hole (the energy source for superradiance is also the 'rotational' energy of the black hole). Note that when a rotating hole is irradiated with electromagnetic waves, the amplification factor is rather modest: at most 4.4 per cent.

Superradiance manifests itself when a black hole is irradiated with other kinds of waves as well. Thus low-frequency gravitational waves incident on a black hole will also be enhanced. The condition necessary for superradiance of any kind of radiation to arise is always the same: the frequency of waves must be sufficiently low. The amplification factor is found to be different for different radiations. Thus it reaches 138 per cent for gravitational waves, much greater than for electromagnetic radiation.

But let us return to electromagnetic waves.

Imagine that a rotating black hole is surrounded by an artificial sphere reflecting electromagnetic waves. Inside this sphere let there be a minute amount of electromagnetic waves for which the super-radiance condition is satisfied. These waves fall on the black hole, are amplified, and propagate away from it. Having reached the surrounding sphere, they are reflected back to the black hole where amplification is repeated. The process is repeated again and again, so that the energy of the amplified radiation increases, as in an avalanche.

If a hole is made in the reflecting sphere, amplified waves will partly escape through it and our 'device' will work as a generator of electromagnetic radiation, directly transforming the 'rotational' energy of the black hole into electromagnetic radiation.

Assume now that the sphere has no holes and that its walls completely reflect the amplified electromagnetic radiation. The enhancement of the intensity of the electromagnetic energy within the device will then continue to a catastrophic degree when the radiation pressure tears the sphere apart: the system explodes. This invention was called the gravitational 'bomb'.

Note that such gravitational devices are absolutely unfeasible now for generating electromagnetic energy since we are unable to create black holes artificially by superstrong compression of matter, while natural ones, if they exist, lie very far in cosmic space.

Over the brink of the gravitational abyss

So far we have mentioned processes in the space around a black hole. Let us turn now to the most intriguing and breathtaking aspect: let us try to approach the edge of the black hole, the brink of this bottomless abyss (that cannot be filled with anything), and attempt to look into it.

Actually, we know that the word 'look' is unwarranted here. Even if we reach the boundary of a black hole, we cannot see what is going on inside it – unless we continue to plunge in. This is in principle possible, for instance, if we fall freely (in a spaceship) in the gravitational field of the black hole. This falling observer reaches the horizon in a finite time by his clock, and continues to fall further.

We already know that, inevitably, this voyage has the gravest consequences for the astronaut. Indeed, the black hole does not return anything, does not let anything out into the outer space. Whatever the power of the jet engine, the astronaut will never return. Neither will he be able to send a message about his observations (even though our news may reach him). Nevertheless, the voyage is, in principle, possible. What awaits the traveller inside the hole?

Before going on a trip with the astronaut, we recall another, very familiar gravitational phenomenon: the gravitational tidal force. This force arises because any body in a gravitational field has a finite size while gravitational fields are never uniform. Hence, different points of attracted bodies undergo somewhat different gravitational pull.

Let a body be in the gravitational field of a planet. The points of the body closer to the planet will be pulled more strongly than those farther from it. The difference between these gravitational forces, called the tidal force, tends to stretch and break up the body. The steeper the gravitational field changes from point to point, the stronger the tidal force. This 'difference' force is felt both in free fall and at rest. This is in sharp contrast to the gravitational force as such, which is not felt in free fall.

Under ordinary conditions, say in a spaceship on an orbit around the Earth, tidal forces are, of course, negligible and unnoticeable. The same is true for ordinary bodies on the surface of the Earth. However, tidal forces are proportional to body size. Hence, they are appreciable (and very much so!) for the entire Earth in the gravitational field of the Moon. These forces cause tides in the oceans, thus giving the forces their name.

Now let us return to the observer who falls into the black hole. First we place him on the surface of a star undergoing relativistic collapse. The pressure of the stellar matter now offers practically no resistance to the growing gravitational compression, the surface of the star crosses the gravitational radius and continues to shrink further. The process cannot stop and the surface compresses to a point, with density going to infinity, in a short time interval (by the clock of the observer on the surface of the star). Physicists call this state the *singularity*. What are its characteristics?

Overlooking subtleties, we can give the answer to this question as follows: on the way to singularity, tidal gravitational forces tend to infinity. This means that any body (our imaginary observer included) will be torn apart. The same fate awaits any body falling into the black hole after the star has collapsed; this body also reaches singularity. Is it possible to prevent this fall into singularity, once a body sinks below the horizon?

This has been shown to be impossible. The fall into the singularity is unstoppable. The astronaut rapidly plunges into the singularity regardless of his attempts at manoeuvering or the power of rocket engines.

The 'longest' time that the rocket can survive inside the black hole after having crossed the horizon is roughly equal to the time interval that it takes light to cover the distance equal to the black hole size. This duration is very short. For a black hole of ten solar masses, the 'longest' lifetime is only one hundred-thousandth of a second.

In order to survive for so 'long', the spaceship has to perform the following manoeuver. During the fall into the black hole, the engine throttle must be at maximum power so as to nearly stop the ship at the horizon. Then the engines must be shut down and the ship allowed to go into free fall along the radius (from the horizon to the singularity). The duration of this fall is the maximum lifetime. Any attempt by the astronaut to use jet engines and slow down the fall into the black hole or to send the ship into orbital motion will only shorten the time (by the astronaut's clock) of the fall into the singularity.

How is this possible? The reader is bewildered. He agrees that engines are unable to overcome the overpowering gravitational pull in the black hole and stop the spaceship, but surely braking *must* slow down this fall at least a little, make it a bit longer? Moreover, isn't this acceleration of the fall by braking the motion nonsensical?

No, this turn of events is indeed possible inside a black hole. The point is that the astronaut accelerates his rocket (we denote it by A) with respect to the freely falling one (rocket B). We have mentioned earlier that time flows more slowly in an accelerated rocket. This factor becomes decisive in a black hole. Rocket A falls into the singularity anyway but the clock in it is slowed down considerably from the viewpoint of rocket B, so that the entire fall takes less time by the clock in A. Being slower, the A clock 'ticks away' a smaller number of seconds (or fractions of a second), that is, this clock demonstrates that the fall was shorter! A paradox indeed.

Let us return to the problem of gravitational tidal forces. We will compare the tidal forces acting on an astronaut in a spaceship orbiting the Earth and in one falling into a black hole.

In the former case, tidal forces extend the astronaut's body to an absolutely negligible degree, corresponding to a pressure of one-ten-billionth of one atmosphere.

In the latter situation, tidal forces are already tremendous at the black hole boundary. It is found that the smaller the mass and size of the black hole, the greater the tidal forces at the horizon. For a black hole of a thousand solar masses, tidal forces correspond to a pressure of one hundred atmospheres. The human body cannot withstand such loads. Tidal force at the horizon is even greater for smaller black holes.

Hence, if the black hole mass is less than a thousand solar masses, a man approaching it cannot survive the encounter.

Obviously, even if a spaceship is falling into a very large black hole at whose boundary the tidal forces are not lethal for the human body, it will eventually plunge into the hole towards the singularity, tidal forces will increase without limit, and any body will be unavoidably crushed. Obviously, no astronaut will penetrate into a black hole of his own free will, unless he intends to commit suicide.

We have discussed this eerie imaginary experiment in order to illustrate the essential feature of the fundamental phenomenon inside a black hole: unlimited growth of tidal forces, ending in the singularity. Why is this so important?

In fact, colossal tidal forces in the neighborhood of the singularity modify the physical laws that were established in less extreme conditions. Some of them will be outlined in the second part of this book. For the time being, it is sufficient to say that both space and time are not only strongly 'curved' in the singularity but that they may lose their continuous nature and split into separate indivisible intervals, or quanta. We will not go into it further here, first, because the reader is likely to be already tired of attempting to digest such unconventional things, and second, because theorists do not yet know well what takes place inside. This is indeed the frontier of gravitation science.

However, what we already know about the 'insides' of black holes is extremely interesting.

This knowledge is the fruit of the large-scale dedicated effort of theorists in many countries of the world.

One of the greatest difficulties has been to find out what goes on inside a black hole in a realistic case, not in an idealized situation. What is the difference between real and idealized situations? Theorists resort to idealizations in order to simplify the equations that are to be solved. For example, it was assumed that the contracting star was perfectly spherical, without even the smallest deviations from the spherical shape. Equations for this idealized problem are incomparably simpler than in the general case. They were successfully solved and the 'insides' of the spherical black hole thus born were analyzed. Even after the solution was obtained, it took physicists several decades to thoroughly understand the structure of a black hole's interior.

Actually, no star can be ideally spherical. Deviations from sphericity are enhanced in the course of compression. What happens in this case? Direct methods of solution were powerless here. General solutions were not available. True mathematical sharpness was required to find the answer.

Whenever one reads a paper of this class, the same question comes to mind and is never answered: how could such an unorthodox path to solving the problem be devised? The discoverer of planetary motion Johannes Kepler expressed it very succinctly. He said that people who perceive the meaning of celestial phenomena appear to him nearly as mysterious as the phenomena themselves.

The first success was achieved by the English theorist Roger Penrose. He was able to demonstrate that singularity inevitably appears when a real nonspherical body is compressed inside a mature black hole, that is, compression forms a region of infinite gravitational tidal forces.

Penrose proved that the impossibility of avoiding the formation of a singularity in a black hole is essentially implied by the fact that mapping the entire surface of the Earth on paper cannot be done in such a way that all points that are neighbors on the Earth surface remain to be neighbors on the map. You will remember that Cape Dezhnev and Alaska, being quite close in reality, are often shown at the opposite ends of a map. It is to this familiar fact that Penrose elegantly reduced his proof.

But is it equally inevitable that any body falling into a real black hole also falls into the singularity? Quite a number of theorists tried to tackle this problem. We started to work on it with Andrei Doroshkevich in the middle 1970s, and were later joined by Gürsel from Turkey and Sandberg of the USA. Nowadays these and a number of other specialists have largely exhausted the problem: it was possible to prove that the fall into singularity is inescapable.

I wish to remind the reader that an observer staying outside a black hole has only 'theoretical' knowledge of the events inside it. He cannot obtain any information, any signals from under the horizon of a black hole. The well-known Indian physicist who now lives and works in the USA, Nobel Prize winner Subrahmanyan Chandrasekhar described this situation in very poetical fashion.

For my part, while considering the phenomena associated with event horizons and the impossibility of communication across them, I have often recalled a parable from Nature that I learnt in India fifty years ago.

The parable, entitled 'Not lost but gone before', is about larvae of dragonflies deposited at the bottom of a pond. A constant source of mystery for these larvae was what happens to them, when on reaching the stage of chrysalis, they pass through the surface of the pond never to return. And each larva, as it approaches the chrysalis stage and feels compelled to rise to the surface of the pond, promises to return and tell those that remain behind what really happens, and confirm or deny a rumour attributed to a frog that when a larva emerges on the other side of their world it becomes a marvellous creature with a long slender body and iridescent wings. But on emerging from the surface of the pond as a fully-formed dragonfly, it is unable to penetrate the surface no matter how much it tries and how long it hovers. And the history books of the larvae do not record any instance of one of them returning to tell them what happens to it when it crosses the dome of their world. And the parable ends with the cry . . .

> Will none of you in pity
> To those you left behind,
> Disclose the secret?

Nothing can be simpler nor more complex than a black hole

We have thus made acquaintance with black hole physics, with what happens in the neighborhood of a black hole and what can be expected to happen inside it. The reader will probably agree that black holes are utterly exceptional objects, dissimilar to anything known before them. They are neither bodies in the conventional sense of the word, nor radiation. They constitute holes in space and time, caused by extremely strong curving of space and by changes in the rate of flow of time in the rapidly growing gravitational field.

At the same time, we have shown in the preceding sections that black holes are, in a certain sense, very simple objects. Their properties are completely independent of the properties of the collapsed matter, of all the complexities of material structure, its atomic structure, physical fields in it, of whether the matter was hydrogen or iron, and so on. When a black hole is formed, it is as if all the properties of the body undergoing collapse vanish for an outside observer; they cannot influence either the boundary of the black hole or anything else in the external space; all that remains is the gravitational field, which is characterized by just two parameters: mass and rotation (we have already stated that global electric charge is not typical for celestial bodies). They determine the shape, size, and all other properties of the black hole. We can thus state quite positively that nothing can be simpler than a black hole: a human body is incomparably more complex, not describable by two numbers, as a black hole can be.

This remarkable simplicity of black holes once moved the American physicist Kip Thorn to the following observation: 'It is as though one could deduce every characteristic of a woman from her weight and hair color.'

Nothing, however, can be more complex than a black hole: indeed, human imagination is unable to comprehend the degree to which space is curved and time flow is warped when a black hole is formed. The study of black hole physics extends our knowledge of the fundamental properties of space and time. We will see later that quantum processes occur in the neighborhood of black holes, so that the most intricate structure of the so-called physical vacuum is revealed. Even more powerful (catastrophically powerful) quantum processes take place inside black holes (in the vicinity of the singularity). The experimental discovery of black holes in nature would be of great importance to science. We would be able to study new laws that govern the properties of space and time in strong gravitational fields, and new laws that dictate the motion of matter in extreme conditions. It may be said that black holes are a door to a new, very wide field of study of the physical world.

However, can black holes be regarded as a reality? We have already mentioned that so far any hope of producing them artificially is futile. Fortunately, it was realized that black holes could be formed in the Universe in a natural manner.

4: *The search for black holes*

They must exist

Astronomers' knowledge of stellar evolution leads to an inescapable conclusion: massive celestial bodies at the end of their lifespan must turn into black holes. How is this evolution proceeding and why is the conclusion so definite?

The matter of an ordinary star similar to our Sun is subject to two opposing forces: gravitation, which attempts to squeeze the matter to the central point, and pressure of hot gases that tends to expand the star. The stable state of the star is maintained when the two forces are equal. However, a hot star continuously emits energy from its surface; if this loss were not compensated for, the star would radiate away its thermal energy and begin to shrink. This shrinking does not happen, however, because nuclear fusion reactions in the central region of sufficiently high temperature release tremendous amounts of energy. This nuclear 'burning' first transforms hydrogen, then helium, and then heavier elements: carbon, oxygen, etc. Nuclear reactions constitute the source of energy that stars emit into space.

As time goes on, the store of nuclear fuel in a star is used up. The duration of nuclear 'burning' – this active period in the life of a star – is determined by the rate of energy loss as radiation and by the store of nuclear fuel. Both depend upon the mass of the star. As a result, the lifetime of a star is a function of its mass. Stars of nearly solar mass live for about 10 billion years. Stars of greater mass have shorter life spans. Thus a star of three solar masses exists for one billion years, and one of 10 solar masses, for only 100 million years.

When the store of nuclear fuel is depleted, the star continues to lose energy by radiation and gradually contracts. If its mass does not exceed the solar mass by a factor of more than 1.2, the contraction stops when the star's radius decreases to several thousand kilometers. The density of stellar matter can reach 10^9 g/cm^3. Such stars are called *white dwarfs*. They are familiar objects to astronomers.

Once a white dwarf is formed, the star continues to cool down but its diameter remains almost unchanged. The gas pressure preventing further contraction of a white dwarf is sustained by quantum forces arising between sufficiently closely packed electrons of the plasma matter of the star. In stellar conditions, this pressure is totally independent of the temperature of the matter. Consequently, a white dwarf can completely cool down and turn into a black dwarf without changing its size.

If the stellar mass is greater than 1.2 solar masses, contraction will raise the density to more than 10^9 g/cm^3. At this density, certain nuclear reactions that absorb large amounts of energy come into play. This absorption violates the equality of gravitational and pressure forces and the star begins to contract catastrophically.

One possible outcome of this contraction is the nuclear explosion that we observe as a supernova flaring up. The star throws off its envelope and converts into a so-called neutron star. Gravitational forces compress it so strongly that the density at the center of the star becomes comparable with the nuclear density, 10^{14}–10^{15} g/cm^3.

A neutron star resembles an 'atomic nucleus' of a diameter of about ten kilometers. Nuclear particles in this star (nucleons) are squeezed very tightly to one another. If its mass does not exceed two solar masses, the nucleon gas is capable of withstanding further compression of the star by quantum forces. This is the final state of this star after cooling. In fact, the idea of a 'cool' neutron star is somewhat incongruous in view of typical terrestrial situations. Indeed, heat cannot affect the pressure of such a dense gas, even if the gas temperature reaches hundreds of millions of degrees. It is for this reason that, although astrophysicists often refer to neutron stars as cold, the temperature at the center of the star may reach hundreds of millions of degrees and that on the surface may be a million degrees.

The search for neutron stars was long but unsuccessful. This can be understood very easily. A star only 10 km in diameter, with a surface temperature of a million degrees, can be observed only through the largest telescopes, provided the star is sufficiently close to us. Indeed, the luminous surface of neutron stars is very small and they typically emit a million times less visible light than our Sun. But even if we do observe a neutron star, we still do not know how to distinguish between it and ordinary faint stars.

Attempts were made to identify neutron stars by the gravitational effect they produce on neighbor stars. It would be impossible to recognize a faint neutron star in a close binary since the bright emission of the companion drowns its light. However, neutron stars' masses are mostly the same as those of ordinary stars. Astronomers began to search for binaries with a star of normal mass but very low luminosity. These attempts also proved unsuccessful.

Neutron stars were discovered quite accidentally in 1967 by a group of English astronomers headed by Antony Hewish, 33 years after the theoretical prediction had been made. It was found that close to the surface of neutron stars (they all have very strong magnetic field) there exist active regions that emit well-directed beams of radio waves. An active region rotates together with the surface of the star, and its beam of radio waves acts as a rotating searchlight. The beam scans the sky; when it strikes the Earth, we observe flashes of radio waves arriving at equal intervals corresponding to the rotation period of the star. These were the pulses reported by the British astronomers.

The flashes of radio emission of pulsars (this was the name given to these new cosmic objects) followed with a very short period (about a second or less). This rotation period gives away a star with a diameter of at most several tens of kilometers. Indeed, an equally rapidly revolving star with a diameter of 1000 km (e.g., a white dwarf) would be broken into fragments by centrifugal forces; such fast revolution rates are below the breakdown threshold only in a small neutron star. Thus it has been proved that pulsars are neutron stars.

A pulsar is the final stage in the active life of a star of not too high mass, less than about two solar masses.

In the real Universe, a neutron star is surrounded by interstellar gas. The gas may fall onto the star, get heated up when it strikes its surface and emit X-rays. If a neutron star is in a binary and the atmosphere of its (normal) companion leaks gas, this gas may reach the gravitational field of the neutron star. In this case the gas flux and the intensity of X-ray emission become especially high. Such 'X-ray pulsars' are also found in binary systems.

The existence of neutron stars has thus been proved beyond doubt. Calculations show, however, that if the mass of a star still exceeds the critical value (about two solar masses) even after its nuclear fuel has been depleted, the star has contracted, and the possible processes of shedding the outer shells have been completed, then the tremendous pressure in the superdense nuclear matter becomes insufficient for counteracting the further contraction, so that the transformation into a black hole at the end of the evolution becomes unavoidable.

Sometimes, however, an idea was expressed that, at the final stage of their evolution, massive stars may eject most of their mass into space, while the remnant, whose mass is below the critical value, transforms into a white dwarf or a neutron star. In fact, the majority of astrophysicists came to be of the opinion that this path of evolution would be extremely artificial and rather unlikely. We thus conclude that black holes are inevitably created at the late stages of evolution of massive stars.

Can black holes of nonstellar origin exist in the Universe? The answer is most probably in the affirmative. Later in the book these very interesting and unusual possibilities will be described. Actually, the conclusions on the existence of nonstellar black holes are much less reliable than those on the inevitability of black hole creation at the end of the evolution of massive stars. Furthermore, we will see that at least one black hole has very probably already been discovered by astronomers. For this reason we postpone the discussion of nonstellar black holes and turn to the search for black holes of stellar origin.

How to search for black holes

Apparently, no searches 'in earnest' for neutron stars or black holes were attempted by astronomers before the 1960s. It was tacitly assumed that these objects were too eccentric and most probably were the fruit of theorists' wishful thinking. Preferably, one avoided speaking about them. Sometimes they were mentioned vaguely, with a remark that yes, they could be formed but in all likelihood this had never happened. At any rate, if they existed, then they could not be detected.

Such extraordinary objects violated the picture of the Universe to which astronomers were so accustomed. As far as black holes were concerned, most astronomers expressed doubt openly. Among those who refused to believe in the reality of black holes was the English astronomer Arthur Eddington (1882–1944). His way into astronomy was a classic one. He began as a practicing astronomer at the Greenwich Observatory, and spent much time on studying the statistics of stellar motions. In 1914 he rose to the directorship of the Observatory of Cambridge University, concentrating his research on astrophysics, which was emerging at that moment as a fully-fledged field of science.

His contribution to astrophysics can hardly be overestimated. He was the first to identify the main processes that determine the internal structure of stars; he suggested the extremely important idea that energy from the depths of a star is mostly transported by the slow 'seepage' of light through opaque gas, not by convection resembling water boiling in a saucepan on a heater. Eddington was able to show as early as 1916, when nobody had any notion of nuclear reactions, that gradual contraction accompanied by heating cannot constitute the energy source for stars – this was the reigning dogma – but that there must exist certain profound transformations of matter that Eddington called 'subatomic'. He studied stellar pulsations, the structure of stellar atmospheres, and worked on many other problems of astrophysics.

Eddington was one of the first to comprehend the profoundness and novelty of general relativity. He headed the expedition of 1919 which measured for the first time during a total solar eclipse the deflection of light rays in the gravitational field of the Sun; the result was in complete agreement with the predictions of Einstein's theory. Eddington's scientific merits were recognized everywhere: he was elected President of the Royal Astronomical Society of London, President of the Physical Society of London, President of the International Astronomical Union, and member of numerous academies, including the election as foreign corresponding member of the Academy of Sciences of the USSR.

This was thus the man who was unable to accept the idea that the ultimate fate of a sufficiently massive star is the loss of stability and catastrophic compression. Chandrasekhar recalls that Eddington regarded as impossible a stellar collapse in which gravitation became so strong that it smothered radiation, that is, a black hole is formed.

Chandrasekhar believes that this sharply negative attitude of such an authoritative astronomer delayed the progress of relativistic astrophysics by several decades. How could it happen? Why did a scientist, so progressive and so sensitive to everything new, fail to understand and appreciate such an important idea?

The Soviet astrophysicist Iosif Shklovsky was probably right in suggesting that Eddington had too great a love of stars, to which he devoted his entire life (the life of a single man, a respectable old bachelor). He managed to develop the theory of equilibrium and stability of stars, and had to face this unpleasant catastrophe: relativistic collapse. Eddington's position was that this just cannot be! Nature had to 'invent' a remedy that would protect cosmic matter from such a pitiful fate! Shklovsky had every ground to conclude: 'There is wisdom in saying that our faults are extensions of our merits.'

Even much later, at the end of the 1950s, when I was a student of the astronomy department of Moscow University, not a single professor ever mentioned what became of massive stars when they die. I do not exclude that the notion of death, which is normally avoided in conversation, may have played a role here. Thus the death of stars was a sort of 'forbidden topic for discussion in decent society'.

Moreover, the exotic nature of black holes and the complexity of the subject of the theory of general relativity (the older generation of

astronomers did not make much of it) were additional negative factors.

A number of discoveries made in the 1960s forced the astronomical community to change its attitude towards a large number of processes in the Universe. These were the discoveries of the active nuclei of galaxies and of quasars, which manifested emission power greater than thousands of billions of stars, and the discovery of the primordial microwave background that survived in the Universe from the first moments when it began to expand. In view of this, neutron stars and black holes ceased to seem so very exotic. And finally, as we have already mentioned, neutron stars – pulsars – were discovered. It was the turn of black holes to be found, but how? How could they be detected if they neither emit not reflect light?

In fact, astronomers already had some experience of studying nonemitting objects, such as, for example, dark dust nebulae. They appear as black spots against the background of stars or luminous gas nebulae. However, dust nebulae are objects of gigantic dimensions while black holes of stellar origin are a mere ten kilometers in diameter. Black holes are born of massive stars and thus the closest one to us must lie at a distance of several tens of light years. Correspondingly, the visible angular size of this black hole must be about 0.000 000 01 second of arc; it is absolutely impossible to detect a dark spot of this size.

A black hole has to deflect light rays passing by it. For this effect to be noticeable, the mutual alignment of the light source (a still more distant star), the black hole, and the observer must be so special that it would be hopeless to rely on an accidental realization of this event.

What is left is the possibility of employing the fact that black holes have masses equal to those of large stars but differ from them in not emitting any light. This was the approach of the Soviet astrophysicists Guseinov and Zel'dovich in 1964. They suggested searching for black holes among binary stellar systems, assuming that in some systems one star is a normal luminous star while its companion is a black hole. The two objects must revolve around the common center of mass. However, the black hole is invisible, so that the visible component revolves as if around 'nothing'.

Incidentally, their method of searching for black holes had been proposed by John Michell in 1784 in his paper that predicted the existence of black holes. He wrote:

... since their light could not arrive at us ... we could have no information from sight; yet, if any other luminous bodies should happen to revolve about them we might still perhaps from the motions of these revolving bodies infer the existence of the central ones with some degree of probability, as this might afford a clue to some of the apparent irregularities of the revolving bodies . . ..

Isn't it true that the new is the well-forgotten past?

Let us return now to our century. Of course, no telescope can resolve the orbital motion of a star from a great distance. We can, however, resort to a special method widely used in astrophysics. When a star is moving towards us on its orbit, the lines in its spectrum shift to the blue end, and when it is moving away from us, to the red end. Astronomers are familiar with so-called spectroscopic binaries discovered by the technique described above. When one star of a spectroscopic binary consisting of ordinary stars is moving towards us and the other away from us, the lines shift in the opposite directions. Quite often, periodic displacement of lines in the spectrum of only one star are observed and no spectral lines of the companion are found. Seemingly, a black hole should be suspected in each such case. In most cases, however, the explanation is trivial: the second star is luminous but much fainter than the first; its light is drowned by the radiation of the brighter companion, which makes it invisible.

Soviet astrophysicists proposed to search for extinguished stars among such spectroscopic binaries in which the mass of the invisible companion, calculated on the basis of the motion of the visible one, was greater than the mass of the visible component. This would mean that the invisible component was not an ordinary star but one that had stopped shining. Indeed, if this companion were an ordinary star, it would be the brighter of the two, as the more massive one, and could not be invisible.

However, the faded star can be a white dwarf or a neutron star. In order to identify a black hole among the definitely extinguished stars, it was necessary to prove that the mass of the invisible companion exceeded the critical value (two solar masses). We have already mentioned that the mass of a white dwarf cannot be greater than 1.2 solar masses, and that of a neutron star, greater than 2 solar masses. Hence, if the mass of the faded star was above the critical value and equalled, say, five solar masses, this could only be a black hole.

Following these indications, a search for black holes in spectroscopic binaries was launched in the USSR and later in the USA. These attempts proved unsuccessful. The invisibility of the companion in all 'suspicious' spectroscopic binaries was explainable in conventional terms, without resorting to the black hole hypothesis. The proposed method of searching was too ambiguous because the 'blackness' found by indirect means could almost always be given an alternative interpretation. In fact, 'invisibility' is a poor proof for the existence of something. It sounds like an hoary joke about the title of a thesis: 'The absence of telegraph poles and wire in archeological excavation sites as a proof of the development of radio communications in ancient civilizations'.

It was also found that the method chosen could hardly prove successful, even in principle. It was doomed in view of the very special behavior of stellar evolution in close binaries. Namely, in the course of the evolution the gas flowed from one companion to another, so that the originally more-massive star on the way to collapse passed some of its mass to the less-massive companion. Finally, the visible star was found to have a mass greater than that of the newborn black hole. In a binary of this type, one cannot determine why the companion is invisible: is it an ordinary star that is less luminous than its counterpart (owing to smaller mass) or is it an extinguished star and maybe even a black hole?

It was necessary to find physical phenomena in which the role of a black hole would be active and unambiguous. This phenomenon was found: the falling of gas in the gravitational field of the black hole.

Very large gaseous nebulae have been identified in interstellar space. If a black hole is in such a nebula, the gas will be falling in its gravitational field. Furthermore, the gas carries a magnetic field, so that the falling is accompanied by turbulent flows. As the gas falls, the energy of the magnetic field must transform into heat. 'Heated' electrons moving in a magnetic field emit electromagnetic waves. At the black hole's horizon, effects of the theory of general relativity come into play. Radiation is partly captured into the black hole. The majority of the radiation recorded by a distant observer is emitted from a distance of several gravitational radii. The heated gas is thus radiating energy into surrounding space on the way to the black hole, before falling into it. Could this radiation be sufficient for identifying a black hole at a large distance?

The total amount of radiation (astronomers call this the luminosity) is a function of the amount of falling gas. The luminosity of the gas accreted on a black hole in typical interstellar situations is of the same order of magnitude as that of normal, not exceptionally bright stars. This means that identifying a black hole in this way is very difficult. They lurk among numerous faint stars of the Galaxy. Actually, turbulent motions in the gas falling on a black hole cause rapid oscillations of brightness, the flashes lasting from several hundredths to several ten-thousandths of a second.

A suggestion for looking for black holes in just this way was made at the end of the 1960s by the Soviet astrophysicist Schwarzman. Together with his colleagues, he developed at the special astrophysical observatory of the USSR Academy of Sciences a series of instruments for the experiment called 'Multichannel Analyzer of Nanosecond Pulses of Brightness Variation' (which, in Russian, gave a suitable acronym MANIA; it proved to be symbolic). There followed years and years of dedicated work on designing, implementing, and adjusting the instruments, then trial observations, and finally, the search. Shwarzman followed this road of an experimentalist with nearly maniacal determination. On the way, he carried out interesting observations on a number of celestial objects. Alas, black holes were not discovered in this search.

Has a black hole been identified?

One more approach to looking for black holes was proposed in 1966. To clarify it, first let us answer the following question. The luminosity of the gas falling onto a black hole is relatively low; why?

The point is that the density of interstellar gas is low, so that the amounts of it falling into the hole are small. Can there be conditions in the Galaxy under which much more gas flows towards the black hole?

Yes, such conditions are possible. They are realized, for instance, if the black hole is in a very close binary whose other component is a normal giant star. In this case there will be a powerful flow of gas from the envelope of the normal star into the gravitational field of the black hole companion. We have already discussed this process when mentioning X-ray pulsars in spectroscopic binaries.

The gas in such a binary cannot simply fall onto the black hole: the orbital motion makes it go into rotation and form a disk around the black hole. Friction between gas layers heats the gas to 10^7 degrees (even before it sinks into the black hole). A gas heated to this temperature emits X-rays.

We conclude that black holes must be sought among X-ray sources found in close stellar binaries, along with neutron stars. This prediction was made by Professor Zel'dovich and myself in 1966, soon after the first X-ray sources were discovered. Shklovsky, who also formulated this prediction in 1967, developed a detailed astrophysical picture of processes that must take place in X-ray sources in stellar binaries.

To search for X-ray sources in the sky, X-ray telescopes have to be taken beyond the atmosphere; if observations are to be long, a telescope must be installed on a satellite (a rocket's flight is very short). Such a telescope installed on the UHURU satellite detected a number of X-ray sources in several binaries in 1972. These sources were then carefully studied by instruments installed, for example, on Soviet satellites and spaceships with astronauts on board.

Thus was born the era of X-ray astronomy. This breathtaking branch of science deserves a special book, or several books, but right now our attention is focused on X-ray sources in stellar binaries. Some of the discovered sources changed brightness at a constant period of about a second. These are definitely not black holes: they are rotating neutron stars with a magnetic field, whose magnetic poles do not coincide with the poles of stellar rotation. The gas falls on the magnetic poles along magnetic lines of force, generating the oriented emission of X-rays. Rotation turns these objects into revolving X-ray 'searchlights'. However, a black hole, as we saw, has no active spots on the surface and the 'searchlight effect' cannot arise. Periodic flashes could be produced by clumps of hot gas in the gaseous disk near the black hole, as they rotate in the inner region. However, this period must change rather rapidly because the clumps are not rigidly fixed to anything revolving, and friction will gradually force the clump to move closer to the black hole (the rotation period correspondingly diminishes).

Black holes must therefore be among X-ray sources in binaries that are not pulsars. Note first of all that these sources cannot be ordinary stars. Indeed, for the gas to heat up to a temperature sufficient for emitting X-rays, the gravitational field in which it moves must be very strong. Such fields can be found only near compact (highly contracted) 'dead' stars: white dwarfs, neutron stars, and black holes. How can one single out black holes among the 'dead' stars?

We know that a reliable criterion is the measured mass. If the mass of a 'dead' star is above the two solar masses threshold, the star is a black hole. The mass can be found from the parameters of the orbital motion of stars in the binary. An anlaysis established that at least one of the identified binary X-ray sources had a mass substantially greater than the critical value. This source in the constellation Cygnus is known as Cygnus X-1.

The normal visible star in this binary system is a massive star of about 20 solar masses. The mass of the 'dead' star whose neighborhood is emitting X-rays is about 10 solar masses. This is considerably higher than the critical value. A large number of recent studies improved the reliability of this result. We can thus say, with a good deal of confidence, that the discovery of the first ever black hole in the

Universe has taken place and that this black hole is in the system
containing the Cygnus X-1 source.

Let us have a closer look at the processes occurring in this system.
Its components revolve around the center of mass with a period of
5.6 days. A black hole of about 10 solar masses pulls in the gas from
the atmosphere of the 'normal' giant star of about 20 solar masses.
Orbital motion makes the gas orbit the black hole, and centrifugal
and gravitational forces flatten the gas into a disk.

Owing to the friction of contiguous layers, the gas flowing around
the black hole slowly spirals towards the center. The velocity of
centerward motion is much smaller than that of the orbital motion.
It takes a month for gas parcels to reach the innermost edge of the
disk, the closest to the black hole. As we already know, the orbital
motion becomes unstable here and the gas plunges into the hole.

Friction continues to heat the gas during its entire voyage: the gas
temperature is only several tens of thousands of degrees in the outer
layers, while in the inner layers it reaches more than 10 million
degrees. The total X-ray luminosity of the gas exceeds that of the Sun
(in all spectral regions) by a factor of several thousands. The main
part of the X-ray radiation observed on the Earth arrives from the
innermost parts of the disk, from a region whose diameter is not
more than 200 km. The size of the black hole is about 30 km.

More important evidence that the X-ray radiation from Cygnus X-1
is being emitted from a very small region close to the black hole is
found in the extremely fast random oscillations of the signal,
occurring within several thousandths of a second. If the emitting
object were greater, its brightness could not vary so rapidly.

Such is this extraordinary X-ray source located at a distance of
about six thousand light years from the Earth.

More than ten years have elapsed since the Cygnus X-1 source was
discovered. By now it has been very thoroughly scrutinized. Why
then are we so careful about not going further than announcing the
'probable' discovery of a black hole?

Let me quote two American specialists, Blandford and Thorn:

If this were a routine situation, astronomers would confidently accept the
result. But since man's first discovery of a black hole hangs in the balance,
and since former conclusions are sometimes destroyed by overlooked
systematic errors, the astronomers are being cautious. Until additional,
independent, confirming evidence is found – evidence of a positive rather
than negative 'what else can it be?' nature – they are not willing to conclude
that Cygnus X-1 is definitely a black hole.

Two or three sources similar to Cygnus X-1 have been discovered in recent years; they are candidates for the black hole family – but so far only candidates.

How many black holes are there in our Galaxy? Is there a danger of encountering one of them and falling into this abyss?

It is difficult to answer the first question exactly because we do not know what fraction of massive stars are destroyed completely by a thermonuclear explosion during collapse and what fraction still retains a sufficiently massive core that contracts into a black hole. Most astronomers believe that the Galaxy must contain many millions, maybe billions, of them.

As for the second question, the reader may have answered it already: the probability of accidental collision with a massive dead star is nil. Indeed, stars are spaced at such vast distances that the probability of collision between any two of them is negligibly small. Hence, the probability of collision with a black hole, which is so much smaller than an ordinary star, is many orders of magnitude smaller. Besides, only a very small fraction of all stars in the Galaxy become black holes.

Giant black holes

So far we have discussed the formation of black holes of stellar origin in the Universe. Astronomers have serious grounds for expecting that other black holes, with quite different histories of formation, may also exist.

The reader already knows that very unusual celestial bodies, quasistellar radio sources (quasars), were discovered at the beginning of the 1960s. These objects lie far beyond the boundaries of our Galaxy. They are fantastically powerful emitters of energy: their luminosity sometimes exceeds that of a hundred large galaxies! This in itself is fantastically interesting. But astronomers were virtually dumbstruck when it was established that the quasar emits most of its energy from a region of diameter less than one light year!

For comparison, you may recall that the Galaxy is a hundred thousand light years across.

How could the size of a quasar be evaluated? Are not quasars so far away that they appear as pointlike stars through any telescope which does not allow direct measurements of their size?

Two Soviet astronomers, Efremov and Sharov, and two Americans, Smith and Hoffleit, solved this problem in an indirect way. They found that a quasar's brightness may vary sharply in less than a year. The size of the quasar cannot, therefore, be greater than one light year. Indeed, if it were, the light emitted from its farther regions would arrive with us nearly a year later than that from the closer parts. Even if the quasar's luminosity grew sharply, we would simultaneously record signals of different brightness from different parts of the quasar: bright light from the nearer edge and fainter light from the far edge, since it left the source a year before, at the stage of weak luminosity. In our instruments, the faint and the intensive lights are mixed (and cannot be separated!), so that brightness variations of the entire quasar become smoothed, stretched in time and thus not abrupt.

Therefore, even though a quasar in unbelievably small (only a thousand times greater than the Solar System), it is brighter than ten thousand billion Suns! This just cannot be – such was the unanimous conclusion of astronomers. (I vividly remember this sentence handed down by one very well-known Moscow astronomer from a rostrum, when the entire community was utterly puzzled by the quasar mystery.) However, this 'cannot be' was unacceptable because the astronomers were looking straight at this 'unicorn'.

An avalanche of hypotheses followed, mostly very exotic. Two outstanding astrophysicists, Geoffry and Margaret Burbidge, wrote at the time: '... there are so many conflicting ideas concerning theory and interpretation of the observations that at least 95 percent of them must indeed be wrong.'

By now the only candidate with sufficient claim for the role of the 'main engine' of a quasar was a giant black hole of a mass equal to hundreds of millions of solar masses. The diameter of such a black hole is a billion kilometers.

It has become clear during the past decades that quasars are tremendously active radiant nuclei of large galaxies. Quite often, intensive gas flows are revealed in them. The stars of the galaxy around the quasar are not seen, as a rule, because of the enormous distance and relatively weak luminosity in comparison with that of the quasar. It was also found that the nuclei of many galaxies resemble 'baby quasars' and sometimes manifest violent activity (ejection of gas, variation of brightness, etc.), even if they are less powerful than true quasars. Processes hinting that a weak 'resemblance' of a quasar is also at work are found even in nuclei of quite ordinary galaxies, including our own.

It now seems to be natural that a giant black hole can form at the center of a galaxy. Indeed, the gas contained in the space between stars must be driven by gravitational forces towards the center, forming a huge cloud of gas. The compression of the whole or part of this cloud must lead to the formation of a black hole. Furthermore, central parts of galaxies contain compact stellar clusters composed of millions of stars. In this region stars can be destroyed by tidal forces when they pass at short distances from the already created black hole; the gas of these annihilated stars moves into the neighborhood of the black hole and then sinks into it.

The fall of gas into a supermassive black hole must be accompanied by phenomena similar to those outlined for stellar black holes. However, these processes are incomparably more powerful. Besides, these must be the regions in which charged particles are accelerated by oscillating magnetic fields carried to the black hole by the falling gas.

Under these conditions a rotating black hole may work as a gigantic dynamo. The American physicists Blandford and Znajek demonstrated in the 1970s that if a rotating black hole is placed in an external magnetic field, a powerful electric field is generated in its vicinity. You will remember that the magnetic field is carried here by the interstellar gas flowing towards the black hole. The dynamo will produce current once an external electric circuit is installed. Thus fluxes of charged particles generated near a black hole by the interaction of radiation with the already present particles may constitute such an external circuit.

Taken together, these processes result in the birth of a quasar and in the activity of galactic nuclei.

The above example of a rotating black hole operating as a giant dynamo is very typical.

For a distant observer, the outward appearance of a black hole seems to indicate that it possesses a surface of a certain electric conductivity and certain mechanical properties. Of course no such surface exists! An observer could make sure that it does not, by starting on a voyage into the hole, as we have already explained. Nevertheless, the notion of this apparent surface is very convenient in some cases, just as the notion of a celestial sphere on which celestial objects move is convenient in practical astronomy. There is no such sphere but this notion facilitates both programming of observations and calculations of coordinates of celestial bodies.

The notion of the imaginary surface of a black hole – its 'membrane' – was introduced into astrophysics by Kip Thorn and his colleagues; they demonstrated that in membrane framework, complicated mathematical calculations are very often simplified.

But let us turn back to astrophysical aspects.

It is highly probable that supermassive black holes do exist. The French writer Jules Renard once said: 'A scientist is a person who is almost sure of something.' But specifics of astronomy are such that I will shun such conclusions and summarize this section with one sentence. Only further observations may clarify the situation.

5: Black holes and quanta

How empty is the vacuum?

The excitement due to black holes began in astronomy at the end of the 1950s and the beginning of the 1960s. Years passed, and the riddle was considerably clarified. It was understood that the death of a massive star inevitably yielded a black hole; and quasars were discovered, probably containing at the center a supermassive black hole. Finally, the first black hole of stellar origin was identified in the X-ray source of the Cygnus constellation. Physics theorists found order in the baffling properties of black holes, and gradually grew accustomed to these gravitational abysses that can only absorb matter and swell; black holes seemed doomed to eternal existence.

Nothing foreshadowed a new grandiose discovery. Nevertheless, the discovery did come, as a bolt from the blue, and amazed even the old hands of physics.

The message was that black holes are not eternal! They may disappear as a result of quantum processes occurring in strong gravitational fields. We shall begin this story from rather afar, in order to make the gist of this discovery more understandable.

Let us begin with empty space: the vacuum. For physicists, empty space is not at all empty. This is not a pun. It was established quite long ago that the 'absolute' emptiness, the 'nothing-nothing', is impossible in principle. What then is the empty space for a physicist?

The vacuum is what is left after all particles, all quanta of any physical field are removed. The reader (provided he has not recently refreshed his picture of physics) could be expected to remark that *then* nothing will remain. Oh, no, something will. Physicists say that the space will be filled with a sea of unborn, so-called virtual, particles and antiparticles. Virtual particles cannot just be 'removed'. They cannot convert into real particles in the absence of external fields, that is, when no energy is imparted to them.

A particle–antiparticle pair appears at each point of the empty space for only a very short time, after which the two particles again merge and disappear, returning to their 'embryonic' state. Of course, this simplified language creates only a crude image of the quantum processes that take place. The presence of the sea of virtual particle–antiparticle pairs was established long ago by direct physical experiments. We will not discuss this aspect here, otherwise the digression would take us too far from the main topic.

In order to avoid unwanted puns, physicists refer to empty space as the vacuum. We too will use this terminology.

A sufficiently strong or oscillating field (e.g., an electromagnetic field) may trigger the transformation of virtual particles of the vacuum into real particles and antiparticles.

Both theorists and experimenters became interested in such processes quite a long time ago. Let us consider the process by which an a.c. field creates real particles. This is the process that is important in the case of the gravitational field. Quantum processes are known to be unusual, and often constitute a challenge to 'common sense'. Correspondingly, before discussing the creation of particles by alternating gravitational field, we will look at one simple example from mechanics. It will make the further story clearer.

Imagine a pendulum. Its suspension cord passes through a pulley and the length of the suspension can be varied by taking the cord in or letting it out. Push the pendulum. It starts to swing. The period of vibration depends only on the length of the suspension: the longer it is, the longer the vibration period. Now we pull the cord up very slowly. The pendulum length decreases, the period decreases as well, but the range (amplitude) of vibrations increases. Let us slowly play out the cord to the initial position. The period is restored, as well as the vibration amplitude. If we neglect the damping of vibrations in response to friction, the energy carried by vibrations obviously remains the same in the final state: the same as it was before the entire cycle of changing the pendulum length. However, it is possible to vary the pendulum length in such manner that the vibration amplitude is changed when the suspension length returns to the initial value. To achieve this, one has to jerk the cord at twice the pendulum vibration frequency. This is what we do on a swing. We stretch or pull up our legs in unison with the swinging motion, so that the amplitude increases constantly. Of course, the swing can be stopped if our legs are pulled up in 'counter-unison'.

Electromagnetic waves in a resonator can be 'swung' in a similar manner. A resonator is a cavity with mirror walls that reflect electromagnetic waves. If a cavity with mirror walls and mirror piston encloses an electromagnetic wave, we can change the wave amplitude by displacing the piston forwards and backwards at a frequency twice that of the electromagnetic wave. The amplitude, and hence the intensity, of the electromagnetic wave can be enhanced by moving the piston 'in unison' with wave oscillations, or they can be damped by moving in 'counter-unison'. But if the piston is moved randomly – both in unison and in counter-unison – the wave will on average always be enhanced, that is, the energy is 'pumped' into electromagnetic oscillations.

Now let our resonator enclose waves of all possible frequencies. No matter how we move the piston, there is always a wave for which the movements are in unison. The amplitude and intensity of this wave are amplified. The greater the wave intensity, the larger the number of photons (quanta of electromagnetic field) that the wave contains. Therefore, the piston motion changes the shape of the resonator and thereby creates new photons.

Having outlined these simple examples, we return to the vacuum, to this sea of all possible virtual particles. For the sake of simplicity, we will now consider only one species of particles, namely, virtual photons. It is found that a process similar to the oscillation of resonator size discussed above, which in classical physics results in enhancement of the contained oscillations (waves), produces in quantum physics an 'amplification' of virtual oscillations, that is, it transforms virtual particles into real ones. Thus a variation of gravitational field with time must produce photons whose frequency corresponds to the time in which the field changes. As a rule, these effects are negligible because gravitational fields are weak. In strong fields, however, the situation is quite different.

Another example: very strong electric field creates in the vacuum pairs of charged particles (electrons and positrons).

Let us return to black holes from our short digression on the physics of the vacuum. Can particles be created from the vacuum in the neighborhood of a black hole?

Yes, they can. This has been known for a long time, and was held to be nothing sensational. Thus when an electrically charged body is compressed and transforms into a charged black hole, its electric field is so enhanced that it creates electron–positron pairs. Such processes were studied by Markov and his students. In fact, this process is also possible without a black hole, but in that case the electric field must somehow be enhanced to a sufficiently high level. The effect is not at all specific to black holes.

Zel'dovich proved that particles are also created in the ergosphere of a rotating black hole, and subtract some of its rotational energy. This phenomenon is somewhat similar to the process discovered by Penrose and outlined in Chapter 3.

All these processes are caused by fields around a black hole and changes produced in these fields, but they neither reduce the black hole nor diminish the size of the region from which light or other radiation or particles are allowed to escape.

Hawking's discovery

A sensational discovery was made in 1974 by the English theorist Stephen Hawking. A textbook on gravitation, written by the American theorists Misner, Thorn and Wheeler and published before this discovery was made, gives the following characterization of Hawking's work:

In such scope is exhibited not only a considerable insight, depth, and versatility, but also the gift of an extraordinary determination to overcome severe physical handicaps, to seek out and comprehend the truth.

Hawking was able to show that there exists a quantum process of particle creation by the black hole itself, and by its gravitational field, and that this process reduces the mass and size of the black hole.

At a first glance, this is surprising. Indeed, when a black hole is being formed, all processes on a contracting star are rapidly slowed down and 'freeze out' for distant observers; the gravitational field becomes constant everywhere, not changing with time. This field cannot create particles. Hence, if a varying field produces a (very small) number of particles during the formation of a black hole, the flux of these particles from the newborn black hole will rapidly decay, as will all other processes, as the surface approaches the gravitational radius. Hawking's result actually states that this conclusion is wrong, that the flux will not die out but will continue even after the black hole has already formed. Is there a contradiction?

The point is that the field inside the black hole is not at all frozen. No constancy in time is possible there, and everything inside the black hole has to move, to fall on the center. This is the factor leading to the fantastic process discovered by Hawking. The reader will remember that virtual particles under ordinary conditions in vacuum form, for a short time, antiparticle–particle pairs, which then merge and vanish. In the gravitational field of the black hole, one of the particles created in this way may be below the horizon and immediately start to fall centerwards, while the other particle stays beyond the horizon. The latter particle flies away into space and carries part of the energy of the black hole, and hence, of its mass.

There appears, therefore, the quantum radiation of particles by black holes. Actually, this process is usually quite negligible. Hawking's calculations show that a black hole emits radiation as an ordinary body heated to a very low temperature. Thus the emission from a black hole of one solar mass corresponds to a temperature of one-ten-millionth of one degree. Of course, this radiation is negligibly small. The wavelength of emitted photons equals the size of the black hole: 10 km. Energy loss to this radiation can be completely ignored.

Energy gain due to the rarest atoms in interstellar space and to minute fluxes of light that propagate through the Universe and fall into such a black hole is much greater in the real conditions of today's Universe than the loss due to radiation. Hence, black holes do not shrink but, in fact, grow. The larger a black hole, the lower its radiative temperature. As a result, quantum radiation of giant black holes is absolutely negligible.

Black holes explode!

Having read the preceding paragraphs, the reader may shrug his shoulders in disbelief: 'This is such an insignificant phenomenon. How could it cause such a storm of astonishment and delight among physicists?'

It did so, first of all because physicists were sure before Hawking's discovery that the static gravitational field outside the black hole cannot in any way create particles. As for the variable field within the black hole horizon, it is 'invisible and impalpable' to distant observers and thus can be safely forgotten. However, it is typical of quantum processes that a particle may show up where classical physics would definitely forbid it. For example, a particle may 'seep' through an energy barrier when its energy is insufficient for it to go over this barrier. Hawking proved that this property of quantum particles produces a qualitatively new effect in the case of black holes: quantum evaporation of black holes. Left alone, free of external influences, black holes slowly disappear, transform into thermal radiation, and slowly shrink in space and time. The principal importance of Hawking's discovery lies in the fact that the notion of the eternity of black holes has been rejected.

But this is only part of the story. The smaller a black hole, the higher the temperature of its radiation.

As the black hole mass diminishes in the course of evaporation, its temperature increases and hence the process of evaporation is intensified. When the mass of the black hole drops to a thousand tonnes, its radiative temperature increases to 10^{17} degrees! Evaporation turns into a fantastic explosion. These last thousands of tonnes, compressed into a microscopic volume, are radiated away, or rather exploded, by the black hole in one tenth of a second. The energy released thereby is equivalent to the explosion of a million hydrogen bombs of one megatonne each. This fantastic firework display wipes out what earlier seemed to be a perpetual gravitational abyss.

Of course, this can happen over a very long period of time. Calculations show that if no external factors are involved, a stellar-mass black hole evaporates and explodes at the end of a period of 10^{66} years. Even astronomers give up when faced with such a long stretch of time.

Nevertheless, these processes may become important in the remotest future of the Universe. We will return to this point in the next part of the book.

Let us turn from the final moments in the life of a black hole to its normal state, and see what particles are emitted in this process.

A black hole creates not only photons but other particles as well. Relatively large black holes of several solar masses have such a low temperature that they cannot generate anything but massless particles. These particles always fly at the velocity of light and have no 'rest mass'. These are photons, electron and muon neutrinos, their antiparticle counterparts, and also the not-yet-discovered gravitons, that is, quanta of gravitational waves. A black hole of typical stellar mass produces copious neutrinos (81 per cent of the entire flux), photons (17 per cent), and gravitons (2 per cent). Different particles are emitted in different amounts because they possess different properties. Mostly neutrinos are emitted because their quantum rotation (what is called *spin* in quantum physics) is minimum ($\frac{1}{2}$), and the fraction of gravitons is the lowest because their spin is maximum (2).

Low-mass black holes have high temperatures. Thus the temperature of black holes of mass less than 10^{17}–10^{16} g exceeds 10^{9}–10^{10} degrees. In addition to the particles listed above, these black holes produce electron–positron pairs. Note that such black holes are a mere 10^{-11} cm in size: a thousand times smaller than the atomic diameter.

Black holes of mass less than 5×10^{14} g can also emit muons and even heavier elementary particles.

Such black holes are smaller than an atomic nucleus. Obviously, such tiny black holes could not be formed in the course of stellar

evolution. However, they could have appeared in the distant past. If such 'primary' black holes of mass less than 10^{15} g were born at the beginning of the expansion of the Universe when the matter was very dense (this is theoretically possible, as was pointed out by Zel'dovich and myself and later proved in detail by Hawking), they would all have evaporated by now. For this reason the process discovered by Hawking is extremely important for cosmology.

If I may allow myself a dream (even though a strictly scientific dream), I can imagine the artificial production of tiny black holes in space (in the very distant future). They would accumulate energy spent on their formation and then would radiate it at a prescribed rate at a given particle energy which is determined by the mass of the black hole. Thus a black hole with a mass of 10^{15} g (the mass of a medium-sized mountain) will emit 10^{17} erg per second for about 10 billion years.

Some things remain unclear in this new phenomenon. For instance, we do not know whether a black hole leaves behind no remnant, or the remnant is a particle of the so-called Planck mass (10^{-5} g). It is not clear whether the process of black hole evaporation is observable in the Universe. As for any experiments with black holes in physicists' laboratories, they sound only like science fiction. But even what we already know makes us reconsider a number of aspects of the evolution of matter in our Universe.

This is the end of our story about holes in time and space. Less than a century ago people were not only unaware of these objects but they could not have imagined them even if a space traveller from our time arrived and tried to describe to them these marvels of nature.

I hope that this story has explained at least in part the unusual popularity that the theme of black holes has enjoyed recently. It will be proper to finish this part of the book with a poem that describes the feelings of a person who faces one of the grandest puzzles in nature, contemplating the vast new world that arises when stars die.

Stars perish in an endless universe.
A star falls into its own decay –
The call of death, –
An end to breath and passion –
into the dark void of oblivion
leaving behind a gulf against the shore
of space and time, a crater of gloom
cocooned within the stellar dust.

This hungry maw of other, new-wrought world
would dwarf the dreaded gap of Dante's wild inferno.
This endless, worldless, spherical abyss
where time and space are madly mixed,
where all roads lead to dire annihilation,
where the black wind roars large and everywhere . . .
. . . a maelstrom of stellar dust, stock-still –
a guard of honour on a crater's verge.

Marina Katys
(Translated by Phil Nix
and Jim Beall)

Part II
To the bounds of infinity

1: The Universe after the explosion

The world we live in

In our journey to the world of black holes, we came across something seemingly impossible: the growing gravitational field virtually warped the properties of space and time and led to amazing physical processes. Now we are on a path towards quite different boundaries; we will go far into the depths of the Universe where we will again find the absolute reign of gravitation. Moreover, here we perceive a fact of immense significance: the observable Universe was born in the Grand Explosion ('Big Bang') that occurred about 15 billion years ago; the explosion was caused by a mysterious singularity, resembling the one hiding in any black hole.

Of course, the desire to comprehend the world in which we live has characterized man ever since people learned to think. The history of the evolution of man's picture of the Universe is both interesting and instructive. Numerous excellent books were devoted to this history, but our aim in this book is quite different.

Let us return to our time and to today's knowledge. If we sometimes have to refer to history, it is to the recent past; only infrequently will we visit the more remote past of the science of the entire Universe, called *cosmology*.

The first thing we encounter when trying to understand the nature of the Universe, is the distribution of celestial bodies in space.

We will be mostly interested in the largest scale accessible to astronomers, and begin with the largest structural units of the Universe: the galaxies.

The reader will remember that our Sun is an element of a huge stellar system that astronomers call the Galaxy with a capital G, and sometimes also *our Galaxy*. The total number of stars in the Galaxy is about a hundred billion.

The predominant part of stars in the Galaxy fill a volume resembling a lens 100 thousand light years in diameter and 12 thousand light years thick. (One light year is the distance covered by light in a year; it equals 10^{13} km.) The interstellar space contains gas and dust that form large clouds. Their total mass comes to only about 5 per cent of the total mass of stars. In addition to this 'main body' of the Galaxy, it also includes a spherical component 5–10 thousand light years in diameter. As a rule, the stars in this spherical system are fainter and older than those in the flattened component.

The young hot stars of the flat system (sometimes referred to as the disk) form spiral arms. These arms begin in the central part of the Galaxy and spiral out to its farthest outskirts.

The spiral arms of our Galaxy led to the entire system being called the *spiral galaxy*. Spiral arms include vast accumulations of gas, the so-called gaseous clouds where young stars are being formed.

Stars and gas of the disk participate in the orbital motion around the center of the Galaxy on nearly circular orbits. Our Sun moves through the Galaxy at a velocity of about 250 km/s and completes one rotation in 200 million years. The stars of the spherical component also move around the center but their orbits are very elongated and lie at arbitrary angles to the plane of the disk.

Such are the structure and scales of the great stellar city, as our stellar system is sometimes called.

Other stellar cities – galaxies with lower-case g – lie beyond our Galaxy. The diameters of most observable galaxies are only slightly smaller than that of ours: they are tens of thousands of light years across and consist of billions of stars.

All these stellar systems lie at distances greater than millions of light years from us. Only the nearest to us and the largest are seen by the naked eye as blurred smears, others are observed only through powerful telescopes. Distances are so large that the light of the stars in them produces only a faint glow. The largest telescopes resolve individual bright stars only in the galaxies closest to us.

Galaxies differ in shape, composition of star population, and in the type of stellar motion. Astronomers classify galaxies into four main types.

Most galaxies are spiral, similar to our own. Some galaxies, however, have no spiral in the lens-shaped disk. Correspondingly, they are known as lens galaxies.

Third, a considerable number of galaxies are not disk-shaped and consist of only a spherical component. They are known as elliptic galaxies because on photographs and through a telescope they appear as elliptic ovals. As a rule, these stellar systems contain little gas and practically no regions where young stars are being born.

The class of irregular galaxies is the smallest. These resemble spiral galaxies in which bright clouds, where young stars cluster, do not form spiral arms but group into irregularly distributed spots. These galaxies often contain large amounts of gas.

Even this cursory outline shows the diversity of the world of galaxies. This diversity is even more striking when we compare masses and sizes of galaxies.

You may remember that our Galaxy contains about 100 billion stars. The largest (elliptic) galaxies may contain ten thousand billion stars. There exist at the same time 'dwarf' galaxies consisting of only a million stars.

What galaxy can be regarded as typical? A relatively large one, like our Galaxy, or a much smaller one?

This question is as difficult to answer as the following one: what city is typical, one as large as Moscow or one much smaller? Indeed, there are dozens of small towns for each large metropolis. The same picture is found in the world of galaxies. There are a large number of dwarfs in each giant system.

What is the distribution of galaxies in space?

It was found that the distribution is extremely nonuniform. Most galaxies are in clusters. The properties of clusters of galaxies are as diverse as those of galaxies themselves. To arrange them in an acceptable order, astronomers devised several classifications. As always in such cases, no single classification can be regarded as complete. For our purposes here, suffice it to say that clusters can be subsumed under two classes: regular and irregular clusters.

Regular cluster often have tremendous mass.They are spherical and consist of tens of thousands of galaxies. As a rule, all these galaxies are elliptic or lens-shaped. One or two giant elliptic galaxies are at the center. A regular cluster closest to us lies in the direction of the Coma Berenices constellation, at a distance of three hundred million light years; its diameter is more than ten million light years. Inside this cluster, galaxies move relative to one another at velocities of the order of a thousand kilometers per second.

The masses of irregular clusters are much less impressive. The number of galaxies in them is smaller by a factor of several tens than in regular clusters, and these are galaxies of all types. Cluster shapes are irregular, with subclusters lying inside large clusters.

Irregular clusters may be quite small, down to tiny groups consisting of several galaxies.

What is observed on scales still greater than galaxy clusters? Are there clusters of clusters of galaxies, that is, their superclusters?

Recently the American astrophysicists Peebles, Gregory and Thomson, and Einasto, Saar and Jēveer in Estonia in the USSR, discovered that the largest-scale nonuniformities in galactic distribution are observed as 'cellular structure'. The 'cell walls' contain many galaxies and clusters of them, while the volume within the cell is empty. The cell size is about 300 million light years, the walls being about 10 million light years thick. Large clusters of galaxies are found at the nodes of the cellular structure. Superclusters are certain fragments of this cellular system. They are often considerably elongated into 'filaments' or 'noodles'. Superclusters comprise here and there giant voids that are almost free of any luminous material. What is the next, higher echelon in this hierarchy?

Here we find a new phenomenon. So far we have encountered more and more complicated systems: small systems formed a larger one, larger systems formed a still larger one, and so forth. The Universe resembled a Russian matreshka wooden doll: a small doll inside a bigger one, which is enclosed in another, still larger, matreshka. What was found was that the Universe has the largest matreshka of them all! No larger system is formed of the large-scale structure of 'noodles' and 'cells', the Universe being uniformly filled (on the average) by these elements. On the largest scale (greater than three hundred million light years), the Universe is found to be uniform, that is, to have everywhere the same properties. This is a very important property and at the same time one of the riddles that nature holds for us. For some unknown reason, relatively small scales reveal large clumps of matter, namely, celestial bodies and their systems that get more and more complicated, but the structure disappears on very large scale. This resembles sand on a beach. Close to the surface, individual grains are resolved, but if one looks from a large distance at a large area, the sand is perceived as a uniform mass.

The uniformity of the Universe was established by observations for distances up to ten billion light years!

We will return to the uniformity puzzle later and now take up the question that in all likelihood worries the reader. How was it possible to measure such staggeringly large distances to galaxies and their clusters, and to operate, with such certainty, with the masses and velocities of galaxies?

'Measurement scale' and other tools of astronomers

We will begin with distances. There can be no doubt that the measurement of distances of millions, even billions, of light years has been a miracle performed by modern science.

The measurement of such distances had been out of the question, say, at the beginning of the century. What, then, were the 'measuring sticks' that made it possible to penetrate regions that lie at such unimaginably large distances?

This was an exceptionally hard road for scientists. The progress to gradually greater distances was achieved step by step. Each next step was always based on the success of the earlier one.

The first serious step was made in the middle of the last century. The distances to three nearby stars were measured at practically the same time in Russia, Germany, and Africa. In principle, the basis of these measurements was the same as in measuring distances on the Earth by using a rangefinder. Rangefinders are now even installed in photographic cameras and are thus familiar to everyone. Essentially, the direction to an object sighted through this instrument is somewhat different when viewed through different windows. If the distance between windows (this is called base length) and the angle between directions is known, the distance to the object is easily calculated by using trigonometric formulas. In the rangefinder, this calculation is performed by an elementary mechanical device. The farther the object, the greater the spacing between the windows must be for the measurement to be reliable. This is known as the trigonometric method. When measuring the distances between stars, astronomers use the diameter of the Earth's orbit around the Sun as the base length. The direction to the star is measured at an interval of half a year from diametrically opposite points of the orbit. Even with this enormous base length, the change in the direction to nearby

stars does not exceed one second of arc, so that measurements
require the highest skill and greatest care.

It was found that even the nearest stars lie at distances greater
than one light year.

More than a century has elapsed since the pioneer measurements
of distances to stars. Despite the tremendous progress in instruments
and measurement techniques, a hundred light years remains the
maximum distance reliably measured by the trigonometric method.

One is still unimaginably far from the boundaries of the Galaxy, let
alone from other galaxies.

The next significant step (rather, leap) up the staircase that leads
far away was made at the beginning of this century; it was
essentially based on employing stars whose luminosity varies syste-
matically, that is, variable stars.

It all began with the work of the American astronomer Henrietta S. Leavitt, who studied variable stars of one of the galaxies closest to us: the Small Magellanic Cloud visible in the Southern skies.

Several years after the study started, Leavitt came across an intriguing fact. Twenty five stars were found to be variable, with luminosity being a strictly periodic function of time. Moreover, the greater the luminosity period, the brighter the star was! Leavitt came to a spectacular conclusion: 'Since the variables are probably at nearly the same distance from the Earth†, their periods are apparently associated with their actual emission of light, as determined by their mass, density, and surface brightness.'

It would be difficult to overestimate the importance of this discovery. It became possible to find the luminosity of a star once the period of luminosity variation was known.

We know that the apparent brightness of a star decreases in inverse proportion to the squared distance to the star. A comparison of the true luminosity of a star with the apparent brightness then gives the distance!

Actually, in order to calculate the distance from the period of brightness variations by Leavitt's data, it is necessary to know the true luminosity of at least one such star.

The first attempt to implement this idea was made by Herzsprung. He understood that the stars described by Leavitt in the Small Magellanic Cloud are nothing other than the familiar stars, called cepheids, of our Galaxy. The brightness of cepheids varies because these stars are pulsating. Now it was necessary to determine the true luminosity of at least one cepheid. This is where serious difficulties began. There is not a single cepheid in the neighborhood of the Sun, such that the distance to it could be reliably determined by the trigonometric method and then the true luminosity found for the known brightness and distance.

Numerous attempts were made to find distances to cepheids in our Galaxy. The first estimate was that of Herzsprung himself. We will not describe here the clever indirect techniques he had used. Let me only remark that both the first and many subsequent attempts were so difficult to perform that the results contained considerable errors. These errors were finally ironed out only at the beginning of the 1960s. In fact, this work is so important (it is The Scale of the Universe that we are after!) that elaboration never stops.

† Because all of them are in the same galaxy, i.e., in the Small Magellanic Cloud.

Once the true luminosity of at least one cepheid with known period of brightness variation had been evaluated, it became possible to determine the distance to any cepheid. Indeed, now the function 'period versus true luminosity' for cepheids had been established. To find the distance to any cepheid, it is sufficient to determine its brightness variation period by observation, obtain the corresponding true luminosity, compare it with the apparent brightness, and calculate the distance. If the cepheid is in a cluster or a galaxy, the result is the distance to this system.

Here cepheids are employed as 'standard candles' whose true brightness is known. The technique as described above was accordingly called the 'standard candle' method.

The role of cepheids in measuring distances is so outstanding that the well-known American astronomer Shapley called them 'the most important' stars.

The true luminosity of cepheids is very high: they are a thousand times brighter than our Sun. Therefore, they are observable from sufficiently large distances, up to 15 million light years. They are thus suitable for measuring distances to the nearest galaxies.

However, we need to measure still greater distances!

Another step has to be made for further progress. It is desirable to identify 'standard candles' that are brighter than cepheids and thus visible from much farther off. Such 'candles' have been found. Galaxies are usually surrounded by stellar clusters that are called globular because of their shape.

When cepheids made it possible for us to find distances to the nearest galaxies, true luminosities of globular clusters around different galaxies were compared. It was found that if one selects the brightest globular cluster at each galaxy, the true luminosity of these brightest clusters is practically the same for all galaxies.

Hence it was possible to use globular clusters as the 'standard candle' that is brighter than cepheids.

This method evaluates distances up to 60 million light years. This is sufficient for 'reaching' the nearest clusters of galaxies. Alas, so far we cannot resolve globular clusters at distances above this limit.

The next step involves a still brighter 'standard candle'. It was discovered that the brightest galaxies in different clusters of galaxies have very nearly the same luminosity: about ten times that of our Galaxy.

These brightest 'standard candles' take us out to billions of light years.

Such is the 'hierarchy of scales' that astronomers use on the way to the depths of the Universe.

How are the velocities of the farthest objects measured?

No doubt, any displacement of stars and other objects, from which the velocity at right angles to the line of sight could be calculated, is immeasurably small at the distances that separate us from the nearest galaxies and even more so from those farther away.

The only velocity that can be measured, and measured simply and reliably, is the velocity of accession or recession of celestial bodies with respect to us. This measurement is based upon taking into account the Doppler effect that we have already mentioned in the book. When a celestial body moves towards us, the light it emits is shifted to the blue end of the spectrum, and when it moves away from us, to the red end. The spectroscopic measurement of these shifts makes it possible to calculate the velocity, or rather its component, along the 'line of sight'. Astronomers refer to velocities found from the Doppler effect as 'line-of-sight velocities'.

Finally, let us look at the masses of galaxies and clusters of galaxies. These can be evaluated by using the law of universal gravitation.

Assume that we observe, say, an elliptic galaxy. Stars in it move at certain velocities with respect to one another. They would scatter all over space were it not for gravitational forces. Gravitation due to the entire mass of the galaxy prevents them from running away. Having measured the relative velocities of stars in a galaxy (this can be done by Doppler techniques) and knowing the size of the galaxy, we can evaluate gravitational forces and, hence, the mass generating them.

In measuring the mass of galactic clusters, a similar approach is used, but instead of the motions of individual stars, one operates with the motion of galaxies in a cluster.

Now the reader realizes, in general, how astronomers obtained the numbers that describe the arrangement of the Universe on a large scale.

Another question arises. What is the motion of galaxy clusters and individual remote galaxies?

The answer to this question was the greatest discovery of twentieth century science. It was found that we live in an expanding Universe. Clusters of galaxies recede from one another, and all the matter in the Universe had been set in the state of expansion by a mysterious Big Bang that had occurred in the very remote past.

The Universe has to evolve

The Universe in which we live has to either expand or contract: this was predicted theoretically by the famous Soviet scientist Aleksandr Freidmann in 1922–4. Friedmann's work was rigorously mathematical, based on Einstein's theory of gravitation. Actually, we need not employ rigorous mathematics to understand the essential point of his discovery. Like all truly great discoveries, it is basically very simple.

You remember why an ordinary star neither contracts nor expands. Gravitational forces are balanced by the force produced by the pressure drop between the dense core of the star and its loose surface layer. In contrast to this, the Universe is uniform on the largest scale, so that no pressure drops are possible. Therefore, gravitation is the only significant force there.

As a result, if the vast masses of the Universe are imagined to be at rest with respect to one another at some moment and to be uniformly distributed, gravitation will immediately set them in motion and the very next moment the matter will start to contract. Gravitation can be balanced out in relatively small systems by the circular motion of bodies on orbits (as in the Solar System) or by random motion of bodies on very elongated orbits (as in elliptic galaxies). However, this is impossible in the all-embracing Universe: some bodies would have to be assigned velocities greater than the velocity of light, in violation of nature's laws.

Friedmann's conclusion was that a stationary universe is impossible. Actually, the Universe does not necessarily have to contract under the influence of the gravitational force. If all masses are first given outward velocities, the Universe expands and the gravitation only slows down the receding bodies. It is thus the result of initial conditions, or rather of the physics of the processes that dictated the initial velocities of masses, whether the Universe will contract or expand. Thus the inevitability of the global evolution of the Universe was theoretically discovered.

This idea was absolutely novel and extremely unusual. A number of different schemes describing the structure of the Universe reigned in science, replacing one another in the history of mankind. One feature was common to all (or nearly all) of them: these were the structural schemes, presenting the eternally unchanging 'mechanism of the Universe's clockwork' but not the development, the evolution, or the maturing of this Universe. The notion of stationarity of the Universe seemed self-evident. Most complex properties could unfold in the Universe but why, from what state, and in what direction was the Universe to evolve?

The idea of the evolution of the entire Universe appeared to be nonsensical, or hardly acceptable to even first-class scientific minds. The great Einstein is one example. The creator of general relativity realized how important his theory was for cosmology. Immediately after he completed the development of general relativity, he tried to find out whether the equations of the theory, when applied to the entire Universe, had a static solution, that is, a solution describing a time-independent state. It appeared obvious to Einstein that he had to look for a static, not evolutionary, model of the Universe. But static solutions could not be found when general relativity equations were applied to the Universe. The idea of a static world was so compelling that Einstein lost faith in his equations and even tried to modify them so as to make them have stationary solutions. We will discuss this attempt later.

What made this idea of static Universe so attractive?

In all likelihood, it agreed with the apparent stationarity, and invariable nature of astronomical bodies, be it the Solar System, stars, clusters of stars, or galaxies. The observed constancy of astronomical phenomena on all scales familiar to people was willy-nilly extended to the Universe as a whole. Aristotle had formulated this attitude very clearly in his treatise *On the Heavens*: 'Throughout all past time, according to the records handed down from generation to generation, we find no trace of change either in the outermost heaven or in any one of its proper parts.'

Today, at the end of the twentieth century, we have grown accustomed to the idea of a changing Universe. We realize that stars and other celestial bodies and their systems only seem to stay unchanged. Man observes them for too short a time to be able to notice their evolution. In fact, stars are born, they live, and then die. Their life spans often reach billions of years. The energy radiated by stars is supplied by nuclear reactions burning in their cores. No store of energy is unlimited; the store of nuclear energy is also finite. Hence, the Sun and the stars appeared in a finite past and had a certain history.

Today we are able to observe violent processes of explosion and evolution in such giant systems as galaxies. The matter in galaxies is gradually transformed in nuclear processes within stars. Hydrogen is converted into helium, and then into heavier chemical elements.

A stationary scenario is thus unacceptable for an astronomical system, provided we consider sufficiently long time intervals. If the model of the Universe had to be built anew, we would have to require that the model would be evolutionary, and that it would specify the epoch in which stars, galaxies, etc., began to form.

But let us return to the story of Friedmann's discovery. The first paper proving that the Universe must evolve was received by the editor of *Zeitschrift für Physik* at the end of June 1922. Einstein was so certain that the equations describing the state of the Universe had to have a static solution that he judged Friedmann's paper to be erroneous. In mid-September, *Zeitschrift für Physik* received Einstein's short note. In Fock's words, Einstein remarked in it,

somewhat haughtily, that Friedmann's results seemed suspicious and that he found there an error; after correction, Friedmann's solution reduces to the stationary one.

Friedmann was informed of Einstein's attitude by his Petrograd colleague Krutkov, who was at the time on a mission abroad. In December 1922, Friedmann wrote a letter to Einstein, where he outlined details of his calculations and presented conclusive proof of correctness of his results. The letter ended with these words:

Should you consider the calculation described in my letter correct, may I ask you to inform the editor of *Zeitschrift für Physik* of your opinion; in this case you may deem it possible to publish an addendum to your earlier note or to copy in the journal a suitable excerpt from my present letter.

The letter reached Einstein and has survived in his archive; it seems, however, that he failed to read it at the time or simply overlooked it, being quite sure that Friedmann was wrong.

It May 1923, Krutkov met Einstein in Leiden, in the house of the well-known Dutch physicist Paul Ehrenfest, and persuaded Einstein in a number of discussions that Friedmann was right. In a letter to his sister in May 1923, Krutkov wrote: 'Victory over Einstein in the argument concerning Friedmann. Petrograd's honor has been saved!'

Immediately after the meeting with Krutkov, Einstein sent the following note to *Zeitschrift für Physik* (I quote here the complete text):

On Freidmann's paper 'On the curvature of space'
My preceding note criticized the above-mentioned paper. Freidmann's letter communicated to me by Mr Krutkov made it clear to me that my critique reflected a miscalculation. I believe that Freidmann's results are

correct and shed new light on the problem. It is found that the field equations allow, in addition to static solutions, also dynamic (i.e. variable in time) centrally symmetric solutions for spatial structure.

Later Einstein continued to emphasize the importance of Friedmann's work for the development of modern cosmology. He wrote in 1931: 'Friedmann ... was the first to start on this road.'

The discovery of the expansion of the Universe

Distant systems of stars (galaxies and clusters of galaxies) are the largest structural units of the Universe known to astronomers. These systems are observed over tremendous distances; it was the study of their motions that produced the observational basis for an analysis of kinematics of the Universe.

The measurement of radial velocities of galaxies was pioneered at the beginning of this century by the American astrophysicist Slipher. Distances to galaxies were not known at the time and heated debates raged among astrophysicists: do these objects lie within our Galaxy or far beyond its bounds? Slipher established that most galaxies (36 out of 41 he observed) are receding, at velocities up to two thousand kilometers per second. Only several galaxies were moving towards us. Later it was found that the Sun revolves around the center of our Galaxy at a speed of 250 km/s, so that the 'accession velocities' of these nearby galaxies simply reflect the fact that the Sun is currently moving towards these objects.

Slipher thus established that galaxies are receding from us. Lines in their spectra were shifted towards the red end. This phenomenon is known as the 'red shift'.

Distances to galaxies were determined in the 1920s.

In 1923, the American astronomer Edwin P. Hubble discovered the first cepheid in one of the closest galaxies in the Andromeda constellation. A year later he discovered more than ten cepheids in this galaxy and twenty-two cepheids in another galaxy in the Triangulum constellation.

Cepheids were discovered in other galaxies as well. Distances to these cepheids and, hence, to the galaxies in which they were found, proved to be much larger than the diameter of our own Galaxy. It had thus been finally established that these galaxies are remote stellar systems similar to ours.

Even the pioneering work on establishing distances to galaxies employed other methods in addition to cepheids. One of the methods involved using the brightest stars of a galaxy as distance indicators.

Apparently, the brightest stars have identical luminosity in our Galaxy and in other galaxies, so that they can serve as 'standard candles' for the determination of distance. Being brighter than cepheids, the brightest stars are seen from greater distances and thus offer a more powerful distance indicator.

A comparison of distances to galaxies with recession velocities (these were determined by Slipher and other astronomers and were only corrected for the motion of the Sun within the Galaxy) made it possible for Hubble to establish in 1929 a spectacular relation: the farther a galaxy lies from us, the greater its recession velocity. It was found that the recession velocities of galaxies and distances to them are very simply related: velocity is directly proportional to distance. The proportionality coefficient is now known as the Hubble constant.

According to Hubble's measurements, the galaxies at a distance of a million light years fly away at a velocity of 170 km/s.

Fifty years have elapsed since Hubble's discovery. The power of astronomical tools has increased dramatically, and new studies have confirmed Hubble's law, that is, the proportionality of the recession velocity of galaxies to the distance separating them from us. It was found, however, that Hubble greatly overestimated the proportionality coefficient.

It happened because there was an error in his evaluation of distances to galaxies. They happened to be reduced by a factor of six to ten. This is not surprising because we have already seen that in order to evaluate very large distances, one has to go up the steps of a long ladder, with errors possible at each step.

The main sources of errors were identified only after 1950 when the largest (at the time) 200-inch telescope of the Mount Palomar observatory was put in operation. In 1952, the American astrophysicist Baade established that cepheids of the type used by Hubble were actually about four times brighter than had been previously believed. This conclusion signified that the distances to the nearest galaxies, evaluated via cepheids, were in fact almost twice as large as earlier calculations suggested. Additional corrections tripled the distances to nearby galaxies. An error at this step of the ladder implied errors at the subsequent steps as well. All measured distances to more distant galaxies also had to be tripled.

Before the distance scale was thus reconsidered, all the nearest galaxies seemed to be substantially smaller than ours. This was strange. After re-evaluation, it became clear that quite a few galaxies are as large as ours and some are even larger. This result supported the belief that scale re-evaluation was correct.

At the end of the 1950s, it was found that further rungs of the ladder taking us deeper into the Universe also contained serious errors. Hubble also made mistakes in determining distances to far-away galaxies in which cepheids are not seen. There were two reasons for this. First, to calculate the visible brightness of very faint stars in other galaxies, comparison must be made with known standard sources. This is a very difficult problem, and errors were indeed revealed in the standard procedure.

The second cause of error was that Hubble had mistaken very bright clouds of ionized hydrogen in distant galaxies for the brightest stars (and employed them as 'standard candles'). Viewed from such distances, these clouds looked like bright star-like points. As a result, the distance scale to far-away galaxies was enlarged further by about 2.2.

When all these factors are taken into account, we find that all distances to the farthest galaxies are greater by a factor of six to ten in comparison with Hubble's estimates. A better evaluation cannot yet be given. The Hubble constant was also reduced by the same factor. According to current data, galaxies at a distance of a million light years recede at velocities of about 25 kilometers per second.

Having made these clarifying remarks, we can return to the principal importance of Hubble's discovery for our understanding of the structure of the Universe.

This discovery demonstrated that galaxies fly away from us in every direction and that the velocity of recession is proportional to the distance.

This fact is likely to be met with disbelief. Why is it that galaxies recede from no other object but our Galaxy? Are we really at the center of the Universe?

This conclusion is wrong. Actually, galaxies recede not only from our Galaxy but from one another as well. If we watched the world from a different galaxy, we would observe just the same picture of recession that we see from our stellar system.

To understand this, imagine two galaxies that move away from us in the same direction, and that the second galaxy lies at twice the distance between us and the first galaxy; correspondingly, the second galaxy recedes at twice the velocity of recession of the first. Imagine that we set our observation post on the second galaxy. It moves away from our original galaxy, and an observer here (who naturally considers himself to be at rest) regards our Galaxy as moving away from him, in the opposite direction, at the same velocity. The first galaxy, halfway between our Galaxy and the second galaxy, moves more slowly, so that an observer on the second galaxy decides that the first one moves in the same direction as our Galaxy but that its velocity is lower. This argument applies to any three galaxies.

Therefore an observer in any galaxy perceives the same pattern of galaxies running away from the point where he is stationed.

We can suggest another model to illustrate the situation. Imagine a homogeneous sphere and enlarge its dimensions, say, to twice the original size, preserving the homogeneity of the sphere. Obviously, the distances between any two points inside the sphere will also be doubled, regardless of the choice of these two points. Hence, wherever we place an observer inside the expanding sphere, he will see the same picture of points receding from him inside the sphere. If the sphere is taken to be of infinite radius, we obtain the situation described above, which is independent of the position of the observer.

The fundamental fact is that galaxies do fly outwards: the Universe is expanding. This is a spectacular confirmation of the conclusion of Freidmann's theory on the nonstationarity of the Universe.

The following question is sometimes posed. Assume that the entire Universe is uniformly filled (on average) with clusters of galaxies. The question would be: 'where' and 'into what' is the Universe expanding?

This question is essentially wrong. The Universe is the totality of everything that exists. There is nothing 'outside' the Universe: neither galaxies, nor any other matter, nor anything at all, be it space or time. There is no vacuum into which to expand. Actually, the Universe needs nothing outside it to allow its expansion. The following example will clearly illustrate this statement.

Let there be an infinite plane with points, representing galaxies, spread uniformly on it. Now we stretch this plane uniformly in all directions, so as to enlarge the distances between the points. The plane was originally infinite; where did it expand to? Obviously, these are the properties of the infinite. If we double it, we again get the same infinity.

Let us forget galaxies and the Universe for a while and devote some time to the notion of infinity, since it is crucial for our concept of the Universe.

The infinity is the subject of the branch of mathematics called *set theory*. Typically, people not looking into this field professionally have a very fuzzy (and naïve) picture of infinity. Intuitively, one is inclined to think that infinity is what we get by indefinitely continuing the count 1, 2, 3 .... If this is so, do we need a theory of the infinite?

Actually, the properties of the infinite go far beyond the indefinite continuation of the sequence 1, 2, 3 .... Moreover, these properties are infinitely more varied and overwhelming than any properties of finite numbers or their groups.

We will describe some of them. Let us begin with a story ascribed to the famous mathematician David Hilbert (as rephrased by a contemporary mathematician).

Imagine a hotel with an infinite number of rooms 'enumerated' by the natural numbers

$$1, 2, 3, \ldots$$

All rooms are occupied. A new guest arrives late at night. 'We are full,' says the receptionist. 'That is of no importance,' objects the hotel manager. 'We will move the guest in room 1 to room 2, the guest in room 2 to room 3, the guest in room 3 to room 4, and so on, and offer room 1 to the new guest.'

A thousand more guests arrive the same night. 'We are full,' says the receptionist. 'No problem,' objects the manager. 'The guest in room 1 goes to room 1001, the guest in room 2 to room 1002, and so on, and the newly arrived guests are to be given the vacated rooms from 1 to 1000.'

The guests barely have time to settle in their rooms when a new wave of arrivals floods the hotel. This time the number of new guests is infinite; we denote them by A_1, A_2, A_3, ... 'We are full', says the receptionist. 'It's all right,' says the hotel manager. 'Move the guest in room 1 to room 2, the guest in room 2 to room 4, the guest in room 3 to room 6, that is, each guest will be asked to move to the room with twice the former number. Now we can offer rooms 1, 3, 5, ... to the guests A_1, A_2, A_3, ...'

This story clearly shows that part of the infinite may be equal to the whole. Indeed, let us write the infinite sequence of even numbers as an infinite row, and place the sequence of guests' numbers as a row below the first:

$$2, 4, 6, 8, \ldots$$
$$1, 2, 3, 4, \ldots$$

Each even number corresponds to one guest's number, and vice versa. Hence, the number of even numbers equals the number of all elements of the natural number series. At a first glance, this conclusion disagrees with our intuition. Indeed, even numbers are only one half of the entire set of natural numbers. This is indeed so for any finite set of numbers. But all this changes when we turn to infinite sets, and a part may be found to equal the whole, as we can clearly see by comparing the two sequences written above.

Other examples given in Hilbert's amusing story point to a number of similar properties.

The above examples may give the impression that all infinities are 'identical', that is, that any infinite set of elements can be enumerated by the infinite set of natural numbers, as we have done with the set of even numbers.

This is definitely not so!

The famous mathematician Georg Cantor proved in the last century that the number of points on a segment of a straight line cannot be counted in any way. They cannot be enumerated by the infinite sequence of natural numbers, assigning one number to each point, regardless of the order in which the points are considered. There will always be at least one point left without a number!

This is not very difficult to understand. Indeed, imagine that we take a segment of unit length and characterize the position of each point by its distance from the left end (chosen as the origin). We write these distances as decimal fractions. In fact, the position of each point is written, generally, as an infinite decimal fraction (a fraction with an infinite number of decimal places after the decimal point). Of course, all decimal places beginning with some place may be zeros in some exceptional cases.

Now imagine that someone has succeeded, in contrast to our statement, in enumerating the points of this segment. We then arrange the decimal fractions that characterize the positions of these points on the segment into a table, in the order of their numbers. The first row has the infinite fraction for the position of the point given number 1; the second row has the infinite decimal fraction for the point given number 2, and so on. Our table may look, for example, like this:

$$0.328\,697\,008\,33 \ldots$$
$$0.919\,671\,384\,52 \ldots$$
$$0.000\,637\,011\ \ \ 4 \ldots$$
$$\ldots\ldots\ldots\ldots$$

Let us show that there is at least one point on the segment that is absent from this list, so that the list is definitely incomplete.

To write the decimal fraction characterizing the position of this point on the segment, we do as follows. As the first decimal place after the decimal point we take any number that differs from the first decimal place in the first row of our table (in our example, this is not 3 but, say, 5). For the second decimal place of our fraction we choose any number not equal to that in the second decimal place of the second row of the table (in our example, not 1); we follow this procedure indefinitely. Clearly we get a fraction that is absent from our list. Indeed, it does not coincide with the first row because it definitely differs from it in the first decimal place, not with the second row because it definitely differs from it in the second decimal place, etc.

The point whose distance is given by this fraction is absent from our infinite list, and, hence, has no number.

One might think that enumeration should begin with this point, and other points are to be enumerated after it. As the Dutch mathematician Hans Freudenthal jokingly remarked, this was the strategy of the man who bet that he would eat 20 potatos. Having eaten 19 of them and feeling unable to chew the last one, the man sighed and complained: 'I should have started with this potato.'

Obviously, if we begin the enumeration with the particular point that originally had no number, we can find in just the same way another point that has no number in the new enumeration.

The reader may be rather tired from the need to follow this unusual construction but it is so important that I wanted to outline it in order to convey at least some feeling of how extraordinary are the properties that we find in the realm of the infinite.

The number of points on a unit-length segment of the line is thus definitely greater than that of natural numbers. Mathematicians say that the *cardinality* of the infinite set of points on a segment of a line is *higher* than that of the set of natural numbers.

Infinities are thus not all the same. Some have greater cardinality, that is, are richer in elements, and some lower.

It may seem that the number of points on the entire straight line is certainly greater than on a unit-length segment, since a segment is only a part of the line. But we are now prudent because we remember that the rule 'a part is smaller than the whole' does not work in the realm of the infinite. Indeed, the infinite sets of points on a line and of its segment have the same cardinality. These are indistinguishable infinities!

Furthermore, the infinite set of points on the entire plane and even in the entire three-dimensional space has the same cardinality as the set of points of a segment of a line. All these infinities are identical. A suspicion may arise that since the set of points of the entire infinite space is not richer than the set of points of a line segment, a set with greater cardinality simply does not exist, that is, that this infinity is the greatest.

This guess is incorrect. Mathematicians are able to construct sets of progressively greater cardinality, that is, richer and richer infinities. In other words, the greatest infinity does not exist; this sequence is also infinite.

But let us proceed no farther than this first step into the world of the infinite. A journey into it may prove to be no less captivating than a voyage through the world of black holes or through the far reaches of the Universe, but it would nevertheless be a path to a different field of human knowledge.

Let us return to the expansion of the Universe. After the infinity stories above, we are no longer surprised that an infinite Universe can expand infinitely, needing for this exposition nothing that lies 'outside' the Universe, nothing of what is not the Universe.

As the infinite number of hotel guests in Hilbert's story could be distributed among even-number rooms only, thus doubling the distance between them, so the distance between galaxies in the Universe can be, say, doubled without going 'outside' the Universe.

Another important question arises, however: what is the reason for the Universe's expansion? What gave galaxies their velocities? The reader remembers that the theory of gravitation cannot answer this question. Galaxies are now moving out by inertia and their velocities are slowed down by gravitation.

We shall return to what caused the Universe to expand in the last chapter of this book.

One concluding remark is necessary here. It is sometimes said that owing to the expansion of the Universe, everything that exists is also expanding: galaxies not only recede from one another but they themselves expand, and so do individual stars, our Earth, and all other bodies. This is patently wrong. The recession of galaxies does not affect in the least the dimensions of individual bodies. Cosmological expansion does not act on gravitationally bound bodies, such as galaxies, stars, or the Earth, just as the expansion of a gas cloud does not affect individual gas molecules. There is no doubt that cosmic objects may expand or contract but these changes are caused by internal factors, that is, processes occurring inside these bodies.

Is the Universe indeed expanding?

The conclusion about the expansion of the Universe did not immediately gain general recognition. The idea of the evolution of the entire world was too grandiose. This idea leads to a number of awe-inspiring and far-reaching consequencies. It implies that in the very remote past, when the expansion began, the Universe looked very different from what we observe today. I mentioned at the beginning of this chapter that this idea met with numerous objections stemming partly from the inertia of scientific thinking and partly from preconceived pseudophilosophical arguments. The notion of an unchanging, static Universe seemed to be so much more habitual and nondisturbing. All this stimulated a number of attempts to give an alternative explanation, not based on the Doppler effect, to the observed 'red shift' in the spectra of remote galaxies. If these explanations were true, one could stop regarding galaxies as receding and the Universe as expanding.

This attitude is excellently reflected in the short witty piece of the Canadian humorist and economist Stephen Leacock, *Common Sense and the Universe*:

> For some twenty-five years past, indeed ever since the promulgation of this terrific idea ... we had lived as best we could in an expanding universe, one in which everything, at terrific speed kept getting further away from everything else. It suggested to us the disappointed lover in the romance who leaped on his horse and rode madly off in all directions. The idea was majestic in its sheer size, but it somehow gave an uncomfortable sensation.

Attempts to 'defend' the stationarity of the Universe are sometimes made even nowadays. But what is the physical reality? Is there a physical process that causes photons to become 'redder' but does not involve the recession of galaxies?

In principle, we know such processes.

For a quantum to 'redden', it must lose some of its energy. This may happen when a quantum collides with electrons of the interstellar medium in its long voyage through the cosmos; in some versions of the hypothesis, it collides with other photons in intergalactic space. In all such hypotheses, an interacting photon not only loses some energy but also changes the direction of propagation. As a result, light quanta propagating right in our direction will gradually start moving on diverging trajectories. The images of remote galaxies would be slightly blurred.

No such effect is observed, and for this reason it cannot explain the 'red shift'. Besides, it would require the intergalactic medium to have fantastically high density and would produce a number of other observable effects.

As another mechanism of photon 'red-shifting', a hypothetical decay of photons was suggested, accompanied by the creation of yet unknown particles. The Soviet physicist Bronstein demonstrated as early as in the 1930s that if such a process did exist (it was later shown that it does not), the probability of photon decay would be inversely proportional to frequency. Hence, the longer the wavelength of a photon, the further it would shift to the red end. Quanta of radiowaves would be expected to shift to the red quicker than visible-light quanta. Precise measurements were carried out in the 1960s of the shift of the radiofrequency line at the wavelength 21 cm. This line is clearly visible in the spectra of cold interstellar gas in many other galaxies.

The red shift of all 30 galaxies observed in the radiofrequency range was found to be the same as the red shift of the visible light emitted by these galaxies.

Consequently, the assumption of red-shifting of quanta as a result of their ageing has to be completely rejected.

The only possible explanation of the cosmological red shift is the Doppler effect caused by the expansion of the Universe.

It must be emphasized again that theorists predicted the nonstationarity of the Universe before it was observed by astronomers, and the discovery of red shift confirmed this prediction. We must be amazed not so much by the phenomena of the red shift and the expansion of the Universe (its nonstatic behavior is a direct corollary of the fundamental laws of physics) as by the tenacity and longevity of conservative beliefs.

If observations found no systematic shift in the spectral lines of galaxies, that is, found no evidence of nonstationarity, this would mean that the laws of gravitation need correction and that some yet-unknown universal force prevents the gravitation from making the Universe nonstatic.

In fact, an attempt to introduce such a force had already been made at the birth of modern cosmology by Einstein, before Friedmann's and Hubble's work. This idea is described in the next chapter.

I must add that some astrophysicists point out that some quasars and galaxies may also have, in addition to the cosmological red shift due to the expansion of the Universe, a red shift due to other causes, such as a strong gravitational field or even other yet unknown processes. In principle, such processes are not impossible. I think, however, that the observational data cited to support these hypotheses are ambiguous and inconclusive and can be explained in a conventional manner.

2: Mechanics of the Universe

The Universe in the past

If the expansion of the Universe is a fact, it signifies that the Universe as we see it today is very different from what it was in the past. If galaxies recede from one another, they had to be practically in contact some time ago, and there had been no individual galaxies at still earlier stages. By dividing the distance between galaxies by their recession velocity we find the time that has elapsed since the recession began. We have mentioned that galaxies separated by a million light years (10^{19} kilometers) recede at a velocity of about 25 km/s. Division gives 13 billion years. The recession velocity of galaxies twice as remote being twice as high, the division gives for them the same time estimate.

Galaxies thus began to 'fly away' about 13 billion years ago. However, we remember that errors are possible in the evaluation of distances to galaxies. Hence, there is an uncertainty in the estimates of time that has elapsed since the beginning of expansion. It can be said that this epoch began 10–20 billion years ago.

Our calculations assumed that galaxies move at constant velocities. In fact, expansion velocities are slowed down by gravitation but this factor does not significantly change the numbers given above.

It is of interest to compare the time since the beginning of expansion with the age of selected objects in the Universe. For instance, the age of the so-called globular clusters in the Galaxy is estimated as 10–14 billion years. Our Earth and the Sun are about 5 billion years old.

We thus find that both the age of our planet and, very probably, of stellar clusters are not much less than the time that has elapsed since the Universe began to expand.

The density of matter in the Universe was, therefore, much higher 10–20 billion years ago, close to the moment when expansion began, than we find it today. Galaxies, stars, and so forth, could not exist as individual objects. The entire matter was in the state of a continuous homogeneous medium. This medium separated into individual blobs much later during expansion, and the blobs then formed celestial bodies. We will return to this process later.

A number of other questions arise. Are the conclusions on the onset of expansion and on the state of tremendous density of the entire matter (this state is called *singular*) reliable, and to what extent? What were the processes that took place in this superdense matter? What was the driving force for the expansion of the Universe? And finally, what was there before the expansion began, before the moment of singularity?

All these questions are undoubtedly extremely important and interesting, and we will discuss them in the remaining part of the book.

The gravitation of empty space

The history of the scientific idea of the gravitation of empty space (of the vacuum, as it is called in the current language of science) reflects the same conflict in the traditional belief in an unchanging universe and of the nonstationary behavior implied by the theory of gravitation.

The law of universal gravitation states that all material bodies attract one another. But does the vacuum gravitate? In modern physics, this question was formulated by Einstein as early as 1917. What is meant by the gravitation of the vacuum? How does this question arise? What physical experiments or astronomical observations lead to this problem? In fact, there were no direct data; rather, it was the absence of data on the motion of galaxies that led Einstein to think about the gravitation of the vacuum.

The situation was as follows. Soon after the completion of general relativity, Einstein used it as the basis of an attempt to construct a mathematical model of the Universe. This attempt preceded Freidmann's work and Hubble's discovery of the red shift in the spectra of galaxies, and the idea of a stationary, unchanging Universe was predominant in Einstein's picture of the world: 'Heavens last from eternity to eternity.' We have seen, however, that the law of gravitation imposes nonstationarity on the Universe.

In order to balance out the gravitational forces and make the world static, one has to introduce repulsive forces that are independent of matter. Using such arguments, Einstein defined a *cosmic force of repulsion* that ensured the stationarity of the world. The acceleration that the cosmic repulsive force was to impart to bodies was thought to depend only on the distance separating them, regardless of their masses.

Einstein was able to show that the repulsive force had to be proportional to the distance between bodies. The proportionality coefficient is known as the *cosmological constant*. This constant would have to be very small for the repulsion to balance the attraction of ordinary matter over intergalactic spaces.

Later in this section we will briefly discuss possible physical causes for the force of repulsion. Suffice it to say here that the cause is rooted in quantum processes in the vacuum.

If repulsive forces do exist in nature, they could be detected, in principle, in high-precision laboratory experiments. In fact, the cosmological constant is so fantastically small that the task of its detection in the laboratory is absolutely hopeless. Indeed, it is easy to calculate that for a body in free fall at the surface of the Earth, the additional acceleration due to repulsive forces would be less by 30 orders of magnitude [*sic*] than the actual acceleration of free fall. These forces are negligibly small in comparison with gravitational forces even on the scale of the Solar System or the entire Galaxy. It would not be difficult to show that the acceleration imparted on the Earth by the gravitation of the Sun is 0.5 cm/s^2. At the same time, the acceleration of cosmic repulsion between the Sun and the Earth is 10^{22} times weaker! Obviously, this repulsion (assuming it exists) does not affect the trajectories of bodies in the Solar System and could be detected only by studying the motion of the remotest observable galaxies.

This is how the cosmological constant, describing the repulsive forces of the vacuum, appeared in Einstein's equations of gravitation. The action of these forces is as universal as that of the forces of universal gravitation, that is, it is independent of the physical nature of bodies it is applied to; it is therefore logical to refer to this action as the gravitation of the vacuum, even though gravitation is usually understood as attraction whereas here we deal with repulsion.

Several years after Einstein's work on general relativity, Freidmann's theory appeared. After this, Einstein felt inclined to think that the cosmological constant should not be forced into the equations of gravitation, since their solution for the entire world can be obtained without this constant anyway.

Any foundation for the hypothesis of cosmic repulsive forces disappeared once the red shift in the spectra of galaxies, proving that the Universe is expanding, had been discovered. Actually, a solution describing the expanding world can also be obtained from equations with a cosmological constant. To do this, it is sufficient to assume that the attractive and repulsive forces do not exactly cancel each other out; the dominating force then results in nonstationary behavior. Freidmann's pioneering work had already pointed to this possibility. Observations of the red shift in Hubble's time were insufficiently accurate for deciding which solution is realized in nature: with or without a cosmological constant. Nevertheless, quite a few physicists looked sourly at the cosmological constant in the equations, since on top of being groundless, it made the theory more complicated. Einstein and many other theorists preferred to drop this constant from the equations of gravitation; Einstein even condemned the introduction of the cosmological constant as the 'gravest error of my life.'

We will show later that what he regarded as his blunder was in fact a first step to comprehending the nature of physical interactions between elementary particles and to perceiving the nature of the vacuum. At the beginning of our century, however, discarding the cosmological constant seemed quite natural.

The cosmologists of the 1930s were not to be hurried into rejecting the cosmological constant. They had serious grounds for retaining it. The reader will recall that pioneering measurements of the Hubble constant gave values overestimated by a factor of about 10. If we were to use this estimate to calculate the time since the onset of expansion, we would obtain merely 1–2 billion years, instead of the correct figure of 10–20 billion years. Two billion years is a very short time. First, it is even shorter than the age of the Earth. A second (and much more important) reason is that the age of stars and stellar systems was wrongly estimated at the time as ten thousand billion years, that is, it was judged to be four orders of magnitude longer than the time since the Big Bang.

Today we know that the time since the onset of expansion was underestimated approximately tenfold, while the age of stars was overestimated by more than two orders of magnitude. From our current standpoint, these two estimates are not in contradiction. In the 1930s, however, this discrepancy was regarded as a serious difficulty.

The cosmological constant was thus a means to make the time since the Big Bang agree with the age of stars. The idea of the universal cosmic repulsion was resuscitated.

Let us see how the introduction of cosmological repulsive forces can drastically change the time-scale of expansion of the Universe.

Assume that the cosmological constant is nonzero. Assume also that the Universe began to expand from a state of very high density. With the density being very high at the initial moment of expansion, the gravitational forces are much stronger than the repulsive forces, since they are proportional to density, and slow down the receding motion.

As expansion continues, density finally decreases to a level at which the gravitational and repulsive forces equalize. At this stage, the world expands without acceleration, at a constant pace. If this pace is very low, the almost perfect equality of attractive and repulsive forces will be maintained for a very long period; therefore, the phase of nearly complete cessation of expansion (known as expansion lag) lasts for a long time. Then the density of matter inevitably diminishes and the gravitational force becomes less than the repulsive force. The world then starts to expand at an acceleration imposed by repulsion. The expansion lag can be made very protracted by adjusting the parameters of the model. This hypothesis assumes that the phase of expansion lag is already over and that now the world is expanding at an acceleration.

The introduction of the cosmological constant thus stretches the scale of expansion of the Universe and can make it agree with the age of stars.

Estimates of Hubble's constant were reconsidered in the 1950s. Before that, at the end of the 1930s, it had been established that the main source of energy in stars was the transformation of hydrogen into helium; the modern theory of stellar evolution was completed in the 1950s. Age contradictions disappeared, the cosmological constant was not needed – for the second time.

The year 1967 saw the beginning of the 'third spring' of the idea of the cosmological constant. By that time, astronomers had discovered and scrutinized strikingly unusual objects: quasars that have been briefly characterized in Part I.

Quasars still hold a large number of mysteries and unsolved problems. Here we point to only two characteristics of quasars. First, their luminosities are tremendous so that they are seen from distances even greater than remote galaxies. The farther a quasar, the fainter its apparent brightness must be to us. At the same time, quasars have to obey the laws of the expanding Universe and thus the greater their distance from us the faster they recede, and the 'red shift' in their spectra must be correspondingly greater.

It was therefore expected that the smaller the apparent brightness of a quasar, the greater its red shift should be.

Observations found nothing of the sort. To explain this, the American scientists Petrossian, Salpeter and Shackers conjectured that one possible reason for the absence of a relation between the apparent brightness of quasars and the red shift in their spectra may be the cosmic repulsive force. We will clarify this.

These American scientists emphasized that as a rule quasars are observed at fantastically large distances, much farther away than the remotest galaxies accessible by telescopes. When we observe quasars with large red shifts, that is, quasars at large distances, we record the light emitted very long ago. If this light left a quasar in the epoch corresponding to the expansion lag predicted by theories with the cosmological constant, then the red shift must be nearly the same for nearer and farther quasars. This happens because the observations refer to the period when the world was hardly expanding.

Indeed, imagine that the light ray leaves a quasar in the epoch of expansion lag. It propagates for a long time in the almost non-expanding Universe, and thus does not redden. While this ray of light is still on the way to us, a ray is emitted from another quasar closer to us; this ray will reach the observer on the Earth in our epoch, together with the former ray. The two rays propagate in the nearly static Universe and suffer no red shift. Later, after the expansion-lag phase ends and the Universe is again expanding, the light from both quasars reddens to the same degree. As a result, the light of the relatively close, and hence bright, quasar, and that of the remote, and hence faint, quasar, will have nearly identical red shifts. This means that many quasars may have similar red shifts in their spectra and very different apparent brightnesses, with no interrelation between these quantities.

The Soviet astrophysicists Shklovsky and Kardashov, pointing to other features of quasar spectra, also argued in favor of expansion involving a prolonged lag in the past (and hence, in favor of a nonzero cosmological constant).

Did lag indeed take place in the past expansion of the Universe? Only new observations could answer this question.

Nearly 20 years have elapsed since this problem was first raised. Numerous new observations of quasars were made. The arguments in favor of an expansion lag began gradually to 'dissolve', in astronomers' jargon. New observations revealed that the absence of a relation between the apparent brightness of quasars and spectral red shifts results from a very wide spread in their luminosities. In contrast to the brightest galaxies in clusters, quasars cannot be treated as 'standard candles', so that no such relation can be expected. The same situation would be found if we took candles of very different true brightnesses: their apparent brightness would in no way characterize the distance from us.

Other arguments in favor of lagged expansion had to be dropped, and with them the need for the cosmological constant. It was dropped for the third time!

One says, however, that once a jinnee is let out of its bottle, it is hard to drive it back. The idea of a nonzero cosmological constant proved to be very viable.

Clearly, even if the cosmological constant is nonzero, it is inevitably very small. Of course, it is extremely difficult to prove by observations that it is exactly zero. So might cosmic repulsive forces be a reality, after all?

This consideration makes physicists think about the nature of such forces. We will discuss this point in detail in the section 'Why is the Universe the way that we see it?' For the time being, note only that as a result of the energy of interaction of virtual particles of the vacuum (this was mentioned in Chapter 5 'Black holes and quanta' in Part I), the empty space may constantly contain nonzero, even if small, energy density. The properties of the vacuum are such that stresses must appear together with energy density (resembling stresses in an elastic body). It can be shown that these stresses create the universal gravitational repulsive forces that we have discussed earlier.

It must be emphasized that the nature of the vacuum remains to a great extent unclear to physicists.

The recent progress in elementary particle physics has led to a fairly reliable conclusion that the forces of gravitation of the vacuum hardly played any significant role in the evolution of the Universe in our epoch and in the greater part of the past. However, their role at the very first moments after the Big Bang could have been decisive, and the properties of the vacuum were very different. We will come to this point later, and suffice it to mention here that the time of the 'fourth spring' of the idea of the cosmological constant may have arrived.

The reader is very likely to be left with a feeling of scepticism towards specialists who find arguments supporting the idea of a gravitating vacuum, then come up with counterarguments, later support it anew, then reject it, and so forth. Isn't this vacillation fatal for the faith in the reliability of scientific research, or in science as a whole? Stephen Leacock, in the pamphlet already cited, outlined a similar situation:

It is not that I venture any disbelief or disrespect toward science, for that is as atrocious in our day as disbelief in the Trinity in the days of Isaac Newton. But we begin to doubt ... So we must pick our little book again, follow science, and wait for the next astronomical convention.

Well, joking aside, science is familiar with situations of this sort. A scientific idea is approached from different sides, at different levels of progress in physics, using tools of gradually higher power. A very complicated problem is attacked many times, until the solution is obtained. As a rule, more profound and more complex problems are unearthed in the process.

The vacuum puzzle belongs among such problems.

The future of the expanding Universe

The cosmological constant is thus unlikely to affect the current expansion of the Universe. Let us assume it to be zero, as Einstein did, and see what happens to expansion in the future.

Gravitation slows down the expansion of the Universe, and the future holds two possibilities. If the deceleration due to gravitation is small, the expansion will last for ever. The distance between clusters of galaxies will grow indefinitely. Gravitational forces in the Universe are determined by the mean density of matter. (The mean density is defined as the density that would be obtained if all celestial bodies, all gas clouds, and all galaxies were uniformly spread over the entire space.) The greater the mean density, the stronger the forces. Hence, sufficiently low mean density implies that expansion will never stop. It is possible, however, that today's density is sufficiently high and that the deceleration of expansion is considerable. In this case the expansion will stop some time in the future, to be replaced by contraction.

The situation is quite similar to that of a rocket accelerated to a certain velocity and required to escape from a celestial body. A velocity of 12 km/s is sufficient for breaking away from the Earth and escaping into space, because it is greater than the escape velocity. This velocity is not sufficient, however, to break free of the surface of Jupiter whose escape velocity is, as we have already mentioned, 61 km/s.

A body ejected from Jupiter at a velocity of 12 km/s will go upward but then fall back to the surface.

Therefore the Universe at its current rate of expansion (its current

Hubble constant) is characterized by a critical matter density that separates the two cases.

Calculations show that this critical density is, on the average, ten hydrogen atoms per cubic meter or the equivalent amount of another element. If the true value of density in the Universe is greater than this level, the current expansion will change to contraction in the future, and if it is less, the expansion is eternal.

What is actually observed?

Unfortunately, this question is not easy to answer. One has to take into account all types of matter existing in the Universe because the gravitational field is generated by them all.

Taking into account the matter in stars, galaxies and luminous gas is a feasible (although difficult) task. There may, however, exist a great deal of barely observable or invisible matter in intergalactic space, which does not or almost does not emit or absorb light.

It is exceptionally difficult to take into account this matter, known as 'hidden' or 'dark'.

As a result, we still do not have the exact and exhaustive answer to the above question.

One reliable estimate can nevertheless be made. If only luminous galaxies are considered, the mean matter density in the Universe comes to one thirtieth of the critical density.

If there were no undetectable forms of matter, the expansion of the Universe would be eternal.

The problem of dark matter

Astronomers have serious grounds for suspecting that intergalactic space may contain large amounts of invisible ('hidden') mass. Invisible haloes of dark matter may even surround individual galaxies.

One of the reasons for this suspicion was found in the results of measuring the masses of galactic clusters. The measurements are carried out as follows.

Regular clusters have symmetric shapes, with the density of galaxies in them decreasing smoothly away from the center. There is, therefore, every reason for assuming that clusters are in an equilibrium state, with the energy of motion of the galaxies being balanced out by the mutual gravitational attraction of all masses in the cluster. As we have already described in the chapter on mass measurements, we can determine the gravitational force and, hence, the total mass of all types of matter in the cluster in this case, because all these types participate in the creation of the gravitational field.

For example, this evaluation for the galactic cluster in the Berenice's Hair constellation gives a mass of 2×10^{15} solar masses.

The cluster's mass can, however, be evaluated in a different manner as well. We can count all galaxies in a cluster and multiply the number by the mass of the average galaxy. The result thus obtained is about ten times less than the figure yielded by the first method.

We conclude that the cluster contains invisible matter between galaxies that creates an additional gravitational field, taken into account in the first but not in the second method.

An analysis of data for other clusters of galaxies gives similar results.

Of course, errors are inevitable when either of the methods is used. However, it is unlikely that errors are sufficiently large to explain the discrepancy in the results. Careful analysis shows that it is extremely difficult to put the 'blame' for a paradoxically large mass in clusters on these errors. These conclusions make one look very seriously at the search for dark matter, not only in galactic clusters but also in intercluster space. What can be said about the form in which the invisible matter may exist? Could it be intergalactic gas? Indeed, the volume of intergalactic space is much greater than that occupied by galaxies. For this reason, even though the concentration of intergalactic gas is much lower than the concentration of gas within galaxies, it could nevertheless constitute gigantic amounts of matter.

Let me emphasize that intergalactic gas is not the only candidate for dark-matter status. Other types of matter could also play this role. We will examine this hypothesis later. Now we will return to interstellar gas and look for ways to detect it.

The reader will remember that the gas in the Universe consists mostly of hydrogen. Therefore, searching for gas in intergalactic space means searching for hydrogen. The gas can be in a neutral or ionized state, depending on physical conditions.

We begin by estimating the possible amount of neutral hydrogen.

If the light emitted by a remote source passes through a gas of neutral hydrogen atoms, these atoms will absorb the radiation at certain frequencies. An attempt to detect the presence of neutral hydrogen in the vast spaces between galactic clusters may be based on this absorption. Far-away quasars are then used as light sources. Several attempts have shown that the amount of intergalactic neutral hydrogen is extremely small. Its mass is at least tens of thousands times less than that of the luminous matter in galaxies.

Therefore, even if intergalactic gas does exist, it must be ionized and heated to a high temperature. An analysis shows that the temperature must be above a million degrees. It is not really surprising that this gas is practically invisible despite its high temperature. The point is that its density is very low, the gas is transparent, and it emits very little in the visible range. Nevertheless, this ionized high-temperature plasma emits fairly intensive ultra-violet (UV) light and soft X-ray radiation.

The hot gas can be identified by its UV radiation. Other methods of searching for hot intergalactic gas also exist.

Unfortunately, all the methods were found to be of insufficient sensitivity. Observations failed to detect hot intergalactic gas. The questions about the amount of this gas and about whether its mean density is higher than the mean density of galactic matter remain unanswered.

Now we turn to the gas in clusters of galaxies. Radioastronomical observations show that clusters contain negligible amounts of neutral hydrogen. However, X-ray telescopes mounted on satellites have successfully detected hot ionized gas in rich galactic clusters. Its temperature was evaluated as 30–100 million degrees, and its total mass may be up to 10^{13} solar masses. This is an impressive figure, although the total mass of the cluster in the Virgo constellation is much higher: it exceeds 10^{15} solar masses. The presence of hot gas in clusters thus cannot be the only answer to the dark-matter problem.

One more aspect of this pressing problem was discovered several years ago.

The idea that galaxies are surrounded by huge massive coronas of faint, practically undetectable, objects has gained more and more advocates in recent years. Possible candidates here are, for instance, low-luminosity stars. The corona mass inevitably affects the motion of dwarf galaxies that are satellites of the master galaxy. It is this effect that is now employed in attempts to observe galactic coronas. It cannot be excluded that when these coronas are taken into account, the estimated masses of galaxies in clusters may change drastically and the dark-mass problem may be solved. At present, however, the galactic coronas conjecture still remains questionable.

We have not yet mentioned exotic candidates for the role of dark-matter carriers, such as neutrinos, gravitational waves, and some other types of matter. We will return to such exotic possibilities in the chapter on the neutrino Universe.

We can summarize the above discussion.

The total mass of luminous matter is not sufficient for its gravitational pull to slow down the expansion of the Universe and make it contract. So far we know too little about dark matter. If it exists, its amount is roughly sufficient to raise the total density of matter in the Universe to the critical, or slightly higher, value.

It is very probable that our Universe will expand infinitely or at least for a very long time in the future.

Curved space

We will see shortly that the actual mean density of the Universe is decisive not only for the future of the Universe but also for its extension. This phrase may have alerted the reader. Can a materialist have two opinions about the extension of the Universe? Isn't it quite clear that the space of the Universe stretches without limit in all directions, to infinity?

Presumably, any other opinion leads to postulating a sort of boundary of the material world, with an immaterial 'something' starting beyond it. Throughout the history of science, the only picture of space acceptable to any spontaneous materialist was the space stretching infinitely in all directions. Arguments proving this were formulated lucidly two thousand years ago by the philosophy genius of ancient Rome Titus Lucretius Carrus in his poem *On the nature of things bounded*:

> The whole universe then is bounded in no direction of its ways;
> For then it would be bound to have an extreme point.
> Now it is seen that nothing can have an extreme point, unless
> There be something beyond to bound it, so that
> There is seen to be a spot further than which the nature
> Of our senses cannot follow it. As it is,
> Since we must admit that there is nothing outside the whole sum,
> It has not an extreme point, it lacks therefore bound and limit.
> Nor does it matter in which quarter of it you take your stand;
> So true is this that, whatever place every man takes up,
> He leaves the whole boundless just as much on every side.

Similar arguments on the infiniteness and boundlessness of space have been faithfully repeated ever since, in the course of centuries.

From today's standpoint, these notions are regarded as naïve. The first blow to the traditional outlook was dealt by the theoretical discovery that it is possible to have geometries that differ from the Euclidean one, taught to pupils at school. The discovery was made by the great mathematicians of the nineteenth century, Nicolai Lobachevsky, Janos Bolyai, Bernhard Riemann, and Carl Friedrich Gauss.

What is non-Euclidean geometry? In terms of planimetry, this is very easy to explain: Euclidean geometry studies the properties of geometrical figures on planar surfaces, while non-Euclidean geometries study figures on curved surfaces, say, a sphere or a saddle-shaped surface. Such curved surfaces contain no straight lines and the properties of geometrical figures that they carry are different from those of planar figures. Straight lines are replaced with lines that trace the shortest paths between points. Such lines are known as geodesics. For instance, geodesics on a sphere are arcs of great circles. Meridians on the surface of the Earth are one example of geodesics. One can draw on a sphere triangles whose sides are geodesics, can draw circles and study their properties. All this is easy to imagine. Difficulties appear when we turn not to a two-dimensional surface but to non-Euclidean three-dimensional space. The properties of prisms, spheres, and other solids differ in this space from those we learn about at school. By analogy to surfaces, we can say that this space is curved. Actually, this analogy can hardly help visualize curved three-dimensional space. We live in three-dimensional space; we cannot jump out of it (since there can not be anything 'beyond space'), so that the question 'Where does our real space curve?' is meaningless. In saying that space is curved we mean that its geometric properties are changed in comparison with those of the flat space in which Euclidean geometry reigns.

The reader will probably remember from the section on black holes that general relativity leads to the conclusion that space in strong gravitational fields is curved and its geometric properties are modified.

The greater the scale that we consider when turning to the vast stretches of the Universe, the larger the mass of matter involved and the stronger the gravitational field. If the scale is large, we have to resort to Einstein's theory and take space curvature into consideration.

At this juncture, we encounter an amazing feature. To understand the essence of the new phenomenon, let us return to curved two-dimensional surfaces.

Take a piece of the plane. If we add to it contiguous parts of the plane of gradually greater size, we finally construct an entire plane that stretches without limit to infinity.

Now let us take a small element of a spherical surface. If it is very small, we will not even notice that it is curved. Add neighboring pieces to it, covering a greater and greater region. Now the curving becomes apparent. Continuing with this procedure, we find that curvature makes our surface close onto itself and form a closed sphere. We have failed to extend the curved surface indefinitely to infinity. The surface has closed in. A sphere has a finite surface area but no boundaries. A flat being crawling on a sphere will never meet an obstacle, or edge, or boundary. Nevertheless, the sphere is not infinite!

This clearly shows that if a surface is closed, it may be unbounded but not infinite.

Let us return to three-dimensional space. We find that its curving may be similar to the curving of a sphere. The space may close onto itself and remain boundless but of finite volume (like a sphere has finite surface area).

A visually clear illustration would be extremely hard to invent, but it is possible, however. Now we understand that the arguments in Lucretius's verse reject the boundedness of space by any barrier but not the finiteness of the volume of space: indeed, space may be unbounded but of finite volume.

Models of the Universe constructed by Friedmann show that this can indeed be the case. All it needs is the mean density in the Universe to be greater than critical. The space is then finite and closed; the model is also called closed.

If the mean density of matter in the Universe equals the critical value, the space geometry is Euclidean. This space is said to be flat. It stretches to infinity in all directions and its volume is infinite.

Finally, if the density of the matter is lower than critical, the geometry of space is again curved. But this geometry is similar not to the geometry on a sphere but to the geometry on a saddle-shaped surface. This space extends just as boundlessly in all directions without getting closed. Its volume is infinite. This is the so-called open Universe model. Is our real Universe open or closed?

You will remember that the mean matter density in space is still unknown, that we do not know whether it is higher or lower than the critical value.

We thus do not know whether our Universe is open or closed.

The idea that the world may be closed, with closed space, is certainly very unusual. Just like the idea of an evolutionary Universe, this idea met with difficulties on the road to recognition. Counter-arguments stemmed partly from the same inertia of reasoning and biased approach, and partly from an inadequate educational level of the proponents of the postulate that only an infinite volume of space is compatible with materialistic attitude. I remember one such heated argument in my student days, when the 6th USSR Conference on Cosmogony assembled in Moscow.

Here is a quotation from the address of one of the philosophers at that conference: 'Indeed, if the Universe is assumed to be finite in space, one immediately faces a number of unanswerable questions: How to imagine a universe with finite volume? What is there beyond it? ...'

You notice that these arguments are much more primitive than in the passage from Lucretius, and are based exclusively on appealing to common sense; we have known for quite a long while that this appeal is a worthless argument in a discussion.

Evidently, a conclusion that space may be closed implies no idealistic connotations.

The postulate fundamental to materialistic philosophy is that matter can exist only in space. 'There is nothing in the world but matter in motion, and matter in motion cannot move otherwise than in space and time' (V. I. Lenin). It is the task of natural sciences to establish specific properties of space, for instance, to determine whether its volume is finite or infinite.

Ginzburg once made a remark that appears very relevant: 'Ideology is not determined by the number of cubic centimeters!'

Such disputes are now in the past and it is for science to work out the true structure of the world.

The curvature of space is dictated by the deviation of matter density from the critical value. The larger the deviation, the greater the curvature. Observations show that even if the matter density does differ from the critical level, the difference is not large and the curvature becomes appreciable only over enormous distances of many billions of light years. The shortest path – the geodesic – in closed space of the Universe is closed like a great circle on a sphere (e.g., like the equator). Gliding along this path, we return to the initial point, just as we arrive at the starting point of a round-the-world journey if we follow the equator.

Future observations may prove that matter density is greater than the critical value and the Universe is thus closed. The volume of the

Universe would then be finite but nevertheless tremendously large; the size of the Universe is colossal. The length of the 'equator', that is, of the geodesic circling the Universe, is definitely not less than several tens of billions of light years and, quite probably, it may be much more.

There are, of course, at least as serious grounds for expecting that the matter density does not exceed the critical value, so that the volume of the Universe is infinite.

We will see in the next section that the difference between the open and closed Universes is not as dramatic as it may seem at the first glance.

Horizon

The Universe began to expand about 15 billion years ago. Therefore, no object in the Universe can be older than 15 billion years, and no source emits light for longer than these same 15 billion years. This limit implies a very important corollary: the apparent horizon. The farther away a galaxy lies, the longer it takes light to reach the observer. The light that is recorded by the observer today left this galaxy in the very distant past. The Universe began to expand about 15 billion years ago. Even the light emitted immediately after the beginning of the expansion propagates only to a finite distance: about 15 billion light years. The points in space at this distance from us form what is called the *apparent horizon*. The regions of the Universe beyond the horizon are, in principle, invisible. Galaxies at greater distances cannot be seen: whatever the power of a telescope, the light from these galaxies simply does not have enough time to reach it from beyond the horizon. The red shift of light increases without limit when we observe objects progressively closer to the horizon. On the horizon, the red shift is infinite. As a result, we can see only a finite number of stars and galaxies in the Universe.

Before the theory of the expanding Universe was developed, attempts to describe infinite space uniformly filled with stars met with an unusual paradox. Here is how it arises. In an infinite universe filled with stars, a line of sight will inevitably meet with the luminous surface of some star. As a result, the entire night skies must shine with the surface brightness of the Sun and stars.

The paradox is known as photometric; a number of first-caliber minds tried to resolve it.

The solution came automatically with the advent of the theory of the expanding Universe. Each observer in the expanding Universe has his apparent horizon. Hence, the observer sees a finite number of stars that are distributed in space quite sparsely. As a rule, the line of sight passes by them and meets no star all the way to the horizon. The night skies between stars are therefore dark. In addition, stars have limited life times.

Owing to the apparent horizon, the difference between a closed and an open world is not very essential for us. In both cases, we observe a limited part of the Universe with a radius of about 15 billion light years. In a closed Universe, light does not manage to travel around the world and return to today's observer; obviously, it is impossible to receive the light of our own Galaxy after it has completed the round trip. The closed Universe does not allow one to see 'the back of one's own head'. Even during the entire period of expansion from the singular state to the moment contraction sets in, light covers only one half of closed space and can complete the round trip only during the contraction stage.

Wherever we place an observer, he has his own apparent horizon. All points of a homogeneous Universe are equivalent. As time goes on, the horizon of each observer widens out and he receives light from new regions of the Universe. During 100 years, the horizon grows by about one-hundred-millionth of its value.

One more remark is needed. In principle, what we see close to the horizon is the matter as it was in the very remote past, when its density was much higher than it is now. Matter had not yet separated into single objects and was opaque to radiation. We will return to this point later.

3: The hot Universe

Physics of the initial stage of expansion

The preceding chapters introduced the reader to the mechanics of the expansion of the Universe. Actually, mechanics is not the only aspect of interest. A great number of processes took place in the Universe at its early moments. We know that at the instant of the Big Bang, 15 billion years ago, the density of matter in the Universe was tremendously high. Hence, the physical processes at that stage were quite dissimilar to those we observe now. These past processes predetermined the current state of the world and, among other things, contained the possibility for the appearance of life.

The physics of processes at the outset of expansion is the focus of extreme attention. However, are we really able to say anything about these processes? This is a legitimate question since we mean the very first moment of expansion, and that moment was 15 billion years ago.

In fact, we really *are* able to infer some things.

The point is that the process during the first seconds of the Big Bang had such important consequences for today's Universe and left 'traces' so obvious, that the processes themselves can be reconstructed.

The most important among them were nuclear reactions among elementary particles, occurring at very high particle densities. Such reactions were possible only at the very beginning of expansion, at enormous densities. Of course, no neutral atoms or even complex atomic nuclei could form at the time, and chemical elements formed much later, as a result of nuclear reactions. There was, however, a still earlier period at which elementary particles themselves were formed. Here one comes to intervals measured in unimaginably short units of 10^{-43} s, when density exceeded 10^{93} g/cm^3. This density is greater by a fantastically large factor than the density of atomic nuclei, which is 'merely' 10^{15} g/cm^3. Such mind-boggling numbers are likely to bring ironic smiles to the faces of readers. Can one hope to glean any information about the processes occurring under conditions that it is absolutely impossible to reproduce in terrestrial laboratories?

Many years ago, when Zel'dovich and I were writing a monograph and working on a classification of processes taking place under such conditions of staggeringly high densities, we recalled a parody written by Arkady Averchenko: 'The history of midianites is dark and unknown but is nevertheless divided by historians into three periods: the first about which we know nothing, the second about which we know almost as much as about the first, and the third that succeeded the first two.'

Physics has come a long way in the 20 years since that time and now we can be sure about quite a few things even in the physics of the formation of elementary particles in the expanding Universe.

We can describe nearly everything about the nuclear reactions that occurred in the interval from the first to the three-hundredth second after the Big Bang, with complete certainty. The point is that nuclear reactions resulted in the formation of chemical elements in the Universe.

Calculation of nuclear reactions makes it possible to predict the chemical composition of the matter of which galaxies, stars, and interstellar gas consist. A comparison of the prediction with observations enables one to identify these reactions and, most importantly, to clarify the physical conditions under which these reactions occurred. Now we will leave the exotic processes at 10^{93} g/cm^3 until later sections and look first at nuclear reactions in the Universe during the first seconds and at their results.

Cold or hot beginning?

In principle, two different scenarios are possible for the conditions under which matter could begin expanding in the Universe. This matter could be either cold or hot. We will see that the nuclear reactions involved lead to drastically dissimilar consequences. Historically, the cold birth scenario was considered first, in the 1930s. Nuclear physics was only just appearing at that moment, and there was no theory for reliably calculating nuclear reactions. It was thus assumed at the time that the matter of the Universe was first in the form of cold neutrons.

Later it was found out that this assumption led to a contradiction with observations.

The situation was as follows. A neutron is an unstable particle. When free, it decays in about 15 minutes into a proton, an electron, and an antineutrino. Hence, as the Universe expands, neutrons inevitably decay, and protons appear. A newborn proton would couple to a surviving neutron and form the nucleus of a deuterium atom (a deuteron). A deuteron would join with another deuteron, and so forth. The reaction of forming progressively more complex atomic nuclei would rapidly proceed until alpha particles (nuclei of helium atoms) formed. Calculations show that more complex atomic nuclei would practically never be formed. The entire matter would thus convert to helium. This conclusion sharply contradicts observations. We know that young stars and interstellar gas mostly consist of hydrogen, not of helium.

Observations of the abundance of chemical elements in nature thus completely reject the hypothesis of the cold beginning of the expansion of the Universe.

In 1948, George Gamov published a paper that suggested a 'hot scenario' of the initial stage of the expansion; this paper was followed by several more written by him and then by his colleagues Ralph Alpher and Robert Herman. It was assumed that the temperature of matter at the onset of expansion was extremely high.

The main objective of the authors of the hot Universe scenario was to analyze nuclear reactions at the start of the cosmological expansion and to obtain the currently observed ratio of abundances of various chemical elements and their isotopes.

Why was it initially assumed that all chemical elements had to be born at the start of the expansion of the Universe? The fact is that it was erroneously decided in the 1940s that the time that had elapsed since the Big Bang was 1–4 billion (10^9) years (instead of the current estimate of 15 billion years). We are aware now that this error was caused by the underestimation of the distances to galaxies and consequent overestimation of Hubble's constant. Comparing this time, $(1-4) \times 10^9$ years, with the age of the Earth, that is, $(4-6) \times 10^9$ years, Gamov and his colleagues concluded that even the Earth and other planets, let alone the Sun and other stars, had to grow from the primary matter and that all chemical elements had formed at the early stage of the expansion of the Universe because they simply could not be formed at any later period.

We know now that the time of expansion of the Universe is 15×10^9 years. The Earth has been formed not from primary matter but from matter that went through the stage of nuclear reactions (nucleosynthesis) in stars. The theory of nucleosynthesis in stars successfully explains the main body of observations of the abundance of chemical elements, assuming that primordial stars grew from matter that mostly consisted of a mixture of hydrogen and helium. The matter of old first-generation stars, enriched in heavier elements, was ejected into space. New stars and planets grew out of this matter. As a result, the need to explain the origin of all elements (including such heavy elements as iron, lead, etc.) at the early stage of expansion disappeared. But the main idea of the hot Universe hypothesis proved to be correct.

A number of researchers pointed out that the content of helium in the stars and gas in our Galaxy is much higher than could be explained in terms of stellar nucleosynthesis. (This will be discussed in detail later.) As a result, the synthesis of helium must have taken place at the early stage of expansion. Nevertheless, the predominant element in the Universe had been, and remains, hydrogen.

The theory suggested by Gamov and his colleagues established that the expanding matter of the Universe turns into a mixture consisting of hydrogen (70 per cent) and helium (30 per cent). This is the material from which stars and galaxies are later formed. What prevents the matter in the hot Universe scenario from turning into helium, as it did in the scenario that started with cold neutron liquid?

The crucial factor is the hot state of the matter. Hot matter contains numerous energetic photons. It also contains protons and neutrons that tend to merge into deuterons. The photons break down the deuterium nuclei formed by paired protons and neutrons, thus interrupting at the very beginning the chain of reactions leading to the synthesis of helium. When the expanding Universe cools down sufficiently (to a temperature of less than a million degrees), a certain amount of deuterium survives and ultimately gives helium. This process will be discussed in detail somewhat later.

The hot Universe scenario gives definite predictions on the content of helium in pre-stellar material. We have already mentioned that the helium abundance must be about 30 per cent of the total mass.

Gamov's hypothesis was not the last attempt to treat the Big Bang. One attempt, made at the beginning of the 1960s, was to modernize the cold Universe scenario; the new version predicted the conversion of all matter not to helium (as in the former version) but to pure hydrogen. It was also assumed that all other elements were generated much later, already in stars.

Originally, the hot and cold Universe models were regarded as attempts to give a detailed explanation of elemental abundances in pre-stellar matter. Attempts at determining which of the theories is true were mostly oriented at an analysis of observational data on abundances of chemical elements. Unfortunately, such observations, and especially their analysis, are extremely complicated and depend on a number of assumptions. If abundances of chemical elements in the Universe were the only test for a theory, the truth could be very hard to unearth. Indeed, it is not easy to decide how much helium and other elements had been synthesized in nuclear processes in stars and how much had been left by the processes occurring in the early Universe.

Fortunately, another method of testing was found. The hot Universe theory gives a most important observational prediction that is a direct corollary of being hot. This is a prediction of the electromagnetic radiation that has survived in the Universe till our epoch from the time in the past when the matter was dense and hot.

The temperature of matter decreases in the process of cosmological expansion, and the radiation also cools off, but nevertheless electromagnetic radiation with a temperature from (in different versions of the theory) a fraction of one degree to 20–30 degrees of the Kelvin scale (20–30 K, or kelvins) must survive until today.

This radiation that must have been left to cool from the ancient epochs of the evolution of the Universe, if the Universe had really been hot at that time, is known as the cosmic microwave radiation background; the Soviet astrophysicist Shklovsky suggested calling it the relict radiation. The electromagnetic radiation at such a low temperature consists of radio waves in the centimeter and millimeter wavelength ranges. Correspondingly, the crucial test of whether the Universe had been hot or cold was a search for such radiation. If found, then the Universe had been hot; otherwise it had been cold.

How the microwave background radiation was discovered

The history of the discovery of the microwave background is highly instructive. The pioneering work of Gamov, Alpher and Herman had already pointed out that the early epochs of the hot Universe would leave behind the radiation that would now have a temperature of about 5 K.

It would seem that this prediction should have attracted the attention of astrophysicists who would pass the news on to radio-astronomers and ask them to try to detect the predicted radiation.

Nothing of the sort happened. Historians of science and specialists are still puzzled by the fact that no one consciously began to search for the microwave background. Before turning to the relevant guesses, let us trace the sequence of actual events that led to the discovery.

In 1960, Bell scientists in the USA constructed a radiowave antenna for studying the reception of signals passively reflected from the Echo I satellite. By 1963, this antenna was no longer necessary for work with the satellite, and two Bell Laboratory radio scientists, Robert Wilson and Arno Penzias, decided to employ it for radioastronomy observations. The antenna was a 20-foot cornucopia-type reflector. Together with the newest receiver, this radio telescope was at that moment the most sensitive instrument in the world for measuring the intensity of radio waves coming from space, from wide areas of the sky. The telescope was meant primarily for measuring the radiation generated in the interstellar medium of our Galaxy. The work was expected to be interesting but more or less routine, one among a large number of radioastronomical observations. At any rate, Penzias and Wilson had no intention of searching for any radiation of cosmological origin, and were totally unaware of the theory of the hot Universe.

The first measurements were conducted at a wavelength of 7.35 cm.

Accurate measurements of galactic radio emission required all possible noises to be taken into account. Radio noise may be of different types. It can originate with the reflection of radio waves in the terrestrial atmosphere and with the radio emission of the Earth's surface. Additional noise is produced by the motion of electric particles in the antenna, in the amplifier's electric circuits, and in the receiver. All possible sources of noise were carefully analyzed and taken into account.

Penzias and Wilson found to their great surprise that in spite of their efforts to eliminate noise, the antenna recorded some radiation of constant intensity regardless of the orientation of the telescope. This noise could not be the radiation from our Galaxy since in this case its intensity would be different for the antenna oriented along the plane of the Milky Way and perpendicular to it. Furthermore, the closest galaxies similar to ours would then also emit the 7.35 cm radiation. No such emission was found.

Two alternative explanations remained: either this was 'noise' of an as-yet unidentified parasitic source, or this was radiation coming from deep in space. There was a suspicion that the noise was connected with the antenna. This gave rise to the 'antenna mystery'. Now I will give the floor to one of the authors of the measurements, Robert Wilson, to describe how he and Penzias tried to locate the source of noise in the antenna.

Thus we seemed to be left with the antenna as the source of our extra noise . . . The most lossy part of the antenna was the small-diameter throat, which was made of electro-formed copper. We had measured similar waveguides in the lab and corrected the loss calculations for the imperfect surface conditions we had found in these waveguides. The remainder of the antenna was made of riveted aluminium sheets, and although we did not expect any trouble there, we had no way to evaluate the loss in the riveted joints. A pair of pigeons was roosting up in the small part of the horn where it enters the warm cab. They had covered the inside wall with a white material familiar to all city dwellers. We evicted the pigeons and cleaned up their mess, but obtained only small reduction in antenna temperature.

For some time we lived with the antenna temperature problem . . .

In the spring of 1965 with our flux measurements finished, we thoroughly cleaned out the 20 ft horn reflector and put aluminium tape over the riveted joints. This resulted in only a minor reduction in antenna temperature. We also took apart the throat section of the antenna, and checked it, but found it to be in order.

Hence, the excess of radiation detected by the radio telescope was not caused by the noise originating in the antenna. It came from the cosmos, at an identical intensity in any direction of observation.

Further events that led to finding the solution to the puzzle were largely accidental. Penzias, when talking to his friend Burke about quite different things, happened to mention the mysterious radiation detected by their antenna. Burke remembered the talk delivered by Peebles, who worked in the laboratory of an outstanding physicist Robert Dicke at Princeton University. Burke said that Peebles had said something about the residual radiation of the early Universe at a temperature around 10 K.

Penzias contacted Dicke by telephone, and the two groups met. Dicke and his colleagues Peebles, Roll and Wilkinson realized that Penzias and Wilson had discovered the primordial radiation of the hot Universe. The Princeton group headed by Dicke was preparing the instruments needed for similar measurements at a wavelength of 3 cm, but had not yet begun actual measurements. Penzias and Wilson had already made their discovery.

Wilson described the subsequent events thus:

We agreed to a side-by-side publication of two letters in the Astrophysical Journal – a letter on the theory from Princeton (Dicke *et al.*, 1965) and one on our measurement of excess antenna temperature from Bell Laboratories. Arno and I were careful to exclude any discussion of the cosmological theory of the origin of background radiation from our letter because we had not been involved in any of that work. We thought, furthermore, that our measurement was independent of the theory and might outlive it. We were pleased that the mysterious noise appearing in our antenna had an explanation of any kind, especially one with such significant cosmological implications. Our mood, however, remained one of cautious optimism.

The two papers were published in the summer of 1965.

The pioneering observations of Penzias and Wilson showed that the temperature of the microwave background was close to 3 K.

Numerous measurements were carried out in subsequent years at various wavelengths, from tens of centimeters to a fraction of a millimeter.

Observations revealed that the spectrum of the microwave background radiation obeys the Planck formula, as would be expected for radiation with definite temperature. This temperature is slightly lower than 3 K.

It was in this accidental manner that Penzias and Wilson made the spectacular discovery of the twentieth century, the discovery proving that the Universe had been hot at the beginning of its expansion. In 1978, they won the Nobel Prize for Physics for this discovery.

Why had the discovery of microwave background not happened earlier?

Let us return now to a problem that belongs to the history of science but that still also excites interest in physicists and astrophysicists. The brilliant American scientist Steven Weinberg wrote in his book *The First Three Minutes: A Modern View of the Origin of the Universe*:

I want especially to grapple here with a historical problem that I find both puzzling and fascinating. The detection of the cosmic microwave radiation background in 1965 was one of the most important scientific discoveries of the twentieth century. Why did it have to be made by accident? Or to put it another way, why was there no systematic search for this radiation, years before 1965?

I remind the reader that the prediction of the radiation at a temperature of several kelvins that fills the Universe was made by Gamov 15 years before the discovery by Penzias and Wilson, at the end of the 1940s and the beginning of the 1950s.

Could it be that sufficiently sensitive radiotelescopes, capable of detecting this radiation, were simply nonexistent? We will see later that this is a rather unlikely reason. Weinberg expresses this opinion too. But the crux of the matter lies elsewhere.

Physics offers numerous examples in which the prediction of a new phenomenon was made long before the discovery could become technically feasible. Nevertheless, if the prediction had solid ground and was important, physicists never let it out of sight. The prediction used to be tested immediately after the technical possibility arose. Weinberg gives an example of the prediction of the antiproton (the antiparticle of the nucleus of the hydrogen atom) in the 1930s. At that period, physicists could not even dream of the equipment needed for its experimental discovery. In the 1950s, however, it became possible, and a special accelerator was constructed in Berkeley to test this prediction.

In the case of the cosmic microwave background, however, radioastronomers knew nothing about it or about the possibilities of detecting it until the middle 1960s.

Why did it happen?

Weinberg gives three reasons. The first is that the hot Universe theory was developed by Gamov and his colleagues to explain the abundances of all chemical elements in nature by their synthesis at the very beginning of expansion. As we mentioned in the preceding section, this hypothesis had to be abandoned since heavy elements were synthesized in stars. Only the lightest elements date back to the earliest moments of expansion. First versions of the theory contained other imperfections as well. Later all shortcomings were corrected for but at the end of the 1940s and in the 1950s the theory as a whole was not yet regarded as quite reliable.

The second reason was the poor communication between theorists and experimentalists. The former had no idea whether the microwave background was detectable by the available observational equipment; the latter never heard that one would be advised to search for this radiation.

The third reason is psychological. It was extremely difficult for physicists and astrophysicists to accept that calculations dealing with the very first minutes after the beginning of expansion could indeed describe the true picture. Indeed, the contrast of durations was too striking: the first several minutes against the tens of billions of years that separate that epoch from ours.

Another reason, which I consider to be the most important, was formulated by Penzias in his Nobel lecture. The point is that neither the first nor subsequent papers by Gamov and his co-workers ever mentioned that the cosmic radiation background could be detected, even in principle. Moreover, it seems that Gamov and the others were of the opinion that the detection was completely unfeasible! Penzias says:

As for detection, they appear to have considered the radiation to manifest itself primarily as an increased energy. This contribution to the total energy flux incident upon the earth would be masked by cosmic rays and integrated starlight, both of which have comparable energy densities. The view that the effects of three components of approximative equal additive energies could not be separated may be found in a letter by Gamov written in 1948 to Alpher (unpublished and kindly provided to me by R. A. Alpher from his files). 'The space temperature of about 5 °K is explained by the present radiation of stars (C-cycles). The only thing we can tell is that the residual temperature from the original heat of the Universe is not higher than 5 °K.' They do not seem to have recognized that the unique spectral characteristics of the relict radiation would set it apart from the other effects.

It so happened that I played my part at the next stage of this story. I started to work in physical cosmology at the beginning of the 1960s, shortly before the microwave background was discovered. I had just completed the postgraduate course at the Moscow University; my science adviser was Professor Zelmanov. My adviser was mostly interested in the mechanics of motion and masses in cosmological models when no simplifying assumptions are made about their uniform distribution. He was less interested in specific physical processes in the expanding Universe. At that time, I knew almost nothing about the hot Universe model.

Not long before the end of my postgraduate term, I was attracted to the following problem. We know how different types of galaxies are studied in different ranges of wavelength of electromagnetic radiation. With certain assumptions about the evolution of galaxies in the past and having taken into account the reddening of light from remote galaxies owing to the expansion of the Universe, one can calculate today's distribution of galactic emission over such wavelengths. In this calculation, one has to remember that stars are not the only sources of radiation, and that many galaxies are extremely powerful sources of radio waves in the meter and decimeter wavelength ranges.

I began the necessary calculations. Having completed the postgraduate term, I joined the group of Professor Zel'dovich; our interest focused mostly on the physics of processes in the Universe.

All calculations were carried out jointly with Doroshkevich. As a result, we obtained the calculated spectrum of galactic radiation, that is, of the radiation that must fill today's Universe if one takes into account only the radiation produced when galaxies were born and stars began to shine. This spectrum predicted a high radiation intensity in the meter wavelength range (such wavelengths are strongly emitted by radio galaxies) and in the visible light (stars are powerful emitters in the visible range), while the intensity in the centimeter, millimeter and some still shorter wavelength ranges of electromagnetic radiation must be considerably lower.

Since the hot and cold Universe scenarios were eagerly discussed in our group (consisting of Zel'dovich, Doroshkevich and myself), the paper that Doroshkevich and I prepared for publication added to the total the putative radiation surviving from the early Universe if it indeed had been hot. This hot Universe radiation was expected to lie in the centimeter and millimeter ranges and thus fell into the very interval of wavelengths in which the radiation from galaxies is weak! Hence, the relict radiation (provided the Universe had been hot!) was predicted to be more intensive, by a factor of many thousands or even millions, than the radiation of known sources in the Universe in this range of wavelengths.

This background could, therefore, be observed! Even though the total amount of energy in the microwave background is comparable with the visible light energy emitted by galaxies, the relict radiation would be in a very different range of wavelengths and thus could be observed. Here is what Penzias said about our work with Doroshkevich in his Nobel lecture:

The first published recognition of the relict radiation as a detectable microwave phenomenon appeared in a brief paper entitled 'Mean Density of Radiation in the Metagalaxy and Certain Problems in Relativistic Cosmology' by A. G. Doroshkevich and I. D. Novikov in the spring of 1964. Although the English translation appeared later the same year in the widely circulated 'Soviet Physics – Doklady', it appears to have escaped the notice of other workers in this field. This remarkable paper not only points out the spectrum of the relict radiation as a blackbody microwave phenomenon, but also explicitly focuses upon the Bell Laboratories 20-ft horn reflector at Crawford Hill as the best available instrument for its detection!

Our paper was not noticed by observers. Neither Penzias and Wilson, nor Dicke and his co-workers were aware of it before their papers were published in 1965; Penzias told me several times that this was very unfortunate.

We are not yet through the chain of missed opportunities that plagued the discovery of the relict radiation.

It is found that the microwave background could have been discovered as early as in 1941. The Canadian astronomer MacKellar was one of the astronomers who discovered molecules in interstellar space. He used the following method of analyzing interstellar gas. If the light of a star is transmitted through a cloud of interstellar gas, the atoms and molecules of this gas absorb the light of stars at very definite wavelengths. As a result, absorption lines of interstellar gas appear in the spectrum.

The positions of lines in the spectrum depend on what element or what molecule caused the absorption and also on the state occupied by atoms or molecules.

In 1941, MacKellar was analyzing absorption lines in the spectrum of the light coming from the star ζ Ophiuchi; the absorption was caused by cyan molecules (compounds of carbon and nitrogen). MacKellar concluded that these lines (in the visible part of the spectrum) could have arisen only if the light was absorbed by rotating cyan molecules. The rotation must be excited by radiation at a temperature about 2.3 K. Of course, neither MacKellar nor anybody else thought at that moment about the possibility that the rotation of molecules is excited by the relict radiation. In fact, the theory of the hot Universe has not even been developed yet!

It was only after the cosmic microwave background had been discovered, that Shklovsky, Field, Hitchcock, Wolf and Thaddeus published their papers, in 1966 and later, which showed that the rotation of interstellar cyan molecules observed in the stellar spectra of the ζ Ophiuchi and three other stars was indeed excited by the relict radiation.

This shows that one manifestation of the microwave radiation background, even though indirect, was observed in 1941: its effect on the rotational state of interstellar cyan molecules.

But even this is not yet the final chapter of the story. Let us return to the question about the technical feasibility of detecting the cosmic microwave background. At what time did this become possible? Weinberg writes: 'It is difficult to be precise about this, but my experimental colleagues tell me that the observation could have been made long before 1965, probably in mid-1950s and perhaps even in mid-1940s.' Is this correct?

In the autumn of 1983, Shmaonov of the Institute of General Physics, with whom I was not previously acquainted, telephoned me and said that he would like to talk to me about things relevant to the discovery of the cosmic microwave background. We met the same day and Shmaonov described how, in the middle of the 1950s, he had been doing postgraduate research in the group of well-known Soviet radioastronomers Khaikin and Kaidanovsky: he was measuring radio waves coming from space at a wavelength of 3.2 cm. Measurements were done with a horn antenna similar to that used many years later by Penzias and Wilson. Shmaonov carefully studied possible sources of noise. Of course, his instruments could not have been as sensitive as those with which the American astronomers worked in the 1960s. Results obtained by Shmaonov were reported in 1957 in his Ph.D. Thesis and published in a paper in the Soviet journal *Pribory i Tekhnika Eksperimenta (Instruments and experimental methods)*. The conclusion of the measurement was: 'The absolute effective temperature of radiation background ... appears to be 4 ± 3 K.' Shmaonov emphasized the independence of the intensity of radiation of direction and time. Errors in Shmaonov's measurements were high and his 4 K estimate was absolutely unreliable, but nevertheless we now realize that what he recorded was nothing other than the cosmic microwave background. Unfortunately, neither Shmaonov himself, nor his science advisors, nor other astronomers who saw the results of his measurements, knew anything about the possibility of the existence of the relict radiation and so failed to pay the results the attention that they deserved. They were soon forgotten. When Doroshkevich and I, having completed our calculations, were calling in 1963 and 1964 on several Soviet radioastronomers with the question 'Do you know of any measurements of the cosmic background in the centimeter and shorter wavelength ranges?', not one of them remembered Shmaonov's work.

It is rather amusing that even the person who made these measurements failed to appreciate their significance, not only in the 1950s – this is easy to explain – but even after the discovery of the cosmic microwave background by Penzias and Wilson in 1965. True, at that time Shmaonov was working in a very different field. His attention turned to old results only in 1983, in response to semi-accidental remarks, and Shmaonov gave a talk on the subject at the Bureau of the Section of General Physics and Astrophysics of the USSR Academy of Sciences. This event took place 27 years after the measurements and 18 years after the publication of the results of Penzias and Wilson.

But even this is not the final chapter. When I was completing the draft of this book, I was informed that measurements by Japanese radioastronomers at the beginning of the 1950s may also have detected the radiation background. This work, as well as that of Shmaonov, went unnoticed then, was not recalled later, and remained practically unknown to the scientific community.

Fate takes unexpected and tortuous turns. Nevertheless, the entire story is very instructive. To hit upon a phenomenon is not yet equivalent to discovering it. One has to realize the significance of the find, and give the correct explanation. A combination of circum-stances and sheer luck do play a role here – no doubt about it. Nevertheless, success does not come by accident. Success requires lots and lots of work, vast knowledge, and persistence in the work itself and in bringing the results to the attention and recognition of others.

Voyage to the remote past

The cosmic microwave background was not born of discrete sources, like the light of stars or radio waves emitted from within galaxies. This radiation background had existed ever since the Universe began to expand. It was contained in the hot matter of the Universe, expanding from the singularity.

If we calculate the total energy density carried by the cosmic microwave background, it is found to be 30 times the energy density of the radiation emitted by stars, radiogalaxies, and other sources taken together. The number of photons of the radiation background, as can be calculated, is about 500 per cubic centimeter of the Universe.

You will remember that the mean density of ordinary matter in the Universe is about 10^{-30} g/cm^3. This means that if we 'spread' the matter uniformly in space, one cubic meter would contain only one hydrogen atom, as hydrogen is by far the most abundant element in the Universe. At the same time, one cubic meter contains about one billion photons of the radiation background.

The quanta of electromagnetic waves, these unusual particles, are thus much more profuse in nature than ordinary matter. There are a billion relict photons for each heavy particle, a proton. If, in addition to hydrogen, we take into account other chemical elements whose nuclei contain not only protons but neutrons as well, our estimate remains practically unchanged since hydrogen is the main element in nature. To recapitulate: there are 10^9 photons for one heavy particle.

We know that today each cubic centimeter of intergalactic space contains about 500 photons that propagate in all directions at the maximum velocity. The energy of each photon corresponds to its frequency. At a temperature of 3 K, most photons have an energy of 10^{-15} erg each. Therefore, each cubic centimeter contains the energy of cosmic microwave background equal to the product of 10^{-15} erg multiplied by 500, that is, 5×10^{-13} erg. According to Einstein's relation, this energy corresponds to the mass of 5×10^{-34} gram. We thus find that today's mass of the microwave radiation background is 5×10^{-34} gram per cubic centimeter.

Recall that the density of ordinary matter is, on average, 10^{-30} gram per cubic centimeter. The mass of matter is thus greater than that of the microwave background by a factor of 2000. Even though the number density of photons is much greater, the ordinary matter greatly dominates the cosmic microwave background in terms of mass. The mass of the latter is negligible.

Let us follow the fate of these two sets of particles into the past.

Neither of these particles was created or vanished in the perceptible past. Certain clarifications are necessary here. The first of them deals with the photons of the cosmic radiation background. Today's Universe is practically transparent to this microwave radiation. Clearly, the relict photons in the current Universe practically never interact with matter and thus their number cannot change. In the very remote past, when matter density was high, the temperature, too, was high. The matter in the Universe was ionized and formed homogeneous plasma. At that moment, it was opaque to radiation. Photons of the radiation background actively interacted with matter. However, the hot matter created as many photons as it absorbed in its depth in any short interval of time. Radiation and matter were, as physicists say, in equilibrium with one another. Consequently, the ratio of a billion relict photons per proton held at that time too.

The second clarification concerns protons.

The Universe was so hot in its remote past, during the very first instant of expansion, that collisions of particles at temperatures above ten thousand billion kelvins created protons and their anti-particles, antiprotons, and also neutrons and antineutrons. We will return to these processes later. As long as we do not look into the exotic first moments, we can assume that radiation background photons and heavy particles are neither created nor annihilated.

Bearing this in mind, we start on a journey into the past. Of course, the number densities of both types of particles were higher than today and as we sink deeper into the past, these densities grow at identical rates. Hence, their ratio remains unchanged: one proton per one billion photons.

There is, however, a tremendous difference between photons and heavy particles. The mass of heavy particles is always the same while the red shift reduces the energy of photons as the Universe expands. If energy changes, the mass of each photon changes as well (its mass is totally related to its energy of motion). Each photon was more energetic in the past, and hence, more massive.

The sum total of one billion photon masses (they grow heavier as we go into the past) becomes equal to the mass of one proton at a certain past moment.

This is the moment when the mass of ordinary matter and that of radiation background in each cubic centimeter become equal. This event occurred when matter density (and radiation density equal to it at that moment) was 10^{-20} g/cm^3, at the temperature of matter and radiation of 6000 degrees. The cosmic background radiation was in visible light then, not in radio waves. Of course, there were no individual celestial bodies yet at that epoch, they formed much later. What was going on even earlier?

At the earlier period, the cosmic radiation background had a mass greater than that of ordinary matter!

Indeed, this had been a most unusual state. We refer to it as the photon plasma epoch.

What we will describe in the lines that follow may resemble frames from a science fiction film. We plan to approach the moment of the beginning of expansion by a truly negligible time interval, a fraction of less than one-hundred-thousandth, and meet most extraordinary processes.

The main fraction of the mass of physical matter in the Universe at the early stages of expansion was that of light; when analyzing this stage, we can safely forget for a while the negligible admixture of ordinary matter to light quanta, although at the present moment this matter is predominant: stars, planets and we ourselves consist of it.

Let us continue our journey to the singularity. For instance, the temperature one second after the start of the expansion was ten million degrees. At earlier moments, the temperature was still higher. When the temperature is so high, elementary particles are created and annihilated. For instance, collision of high-energy photons create electron–positron pairs and electrons colliding with positrons result in annihilation and production of photons, that is, light quanta.

To create an electron–positron pair, one has to spend energy equal to at least the sum of the masses of these two particles times the squared speed of light ($E = mc^2$). Hence, such processes can proceed only at temperatures above ten billion kelvins, when very many quanta of light have such energy. Electron–positron collisions may also create neutrino–antineutrino pairs; likewise, neutrino–antineutrino collisions may produce an electron–positron pair. At still higher temperatures, more massive particles may be created: protons and antiprotons, neutrons and antineutrons, mesons, and so forth.

At temperatures above ten thousand billion kelvins, a great number of particle species (and their antiparticles in equal amounts) coexisted in almost equal abundances; some of them were massive species. As the Universe expanded, the temperature decreased and the energy of particles became insufficient for creating pairs of heavy particles and antiparticles, such as protons and antiprotons. These particles 'died out'.

As the temperature dropped further, a number of meson species died out.

A very important event happened about 0.3 seconds after the start of expansion. The Universe contained at that moment light quanta, electrons and positrons, neutrinos and antineutrinos (for the sake of simplicity, we mention only one species of the neutrino, i.e. the electron neutrino).

At high temperatures, neutrinos and antineutrinos convert into electrons and positrons, and vice versa.

Neutrinos, however, are particles that interact extremely weakly with other objects; even dense matter is virtually transparent to them. About 0.3 seconds after the Big Bang, the entire matter of the Universe, including electrons and positrons, becomes transparent for neutrinos, that is, they stop interacting with the rest of matter. Their number does not change after this, and they survive until today, with their energy reduced by the red shift just as the temperature of electromagnetic radiation quanta is reduced by it.

We conclude that in addition to the cosmic microwave background, the Universe must now also contain relict neutrinos and antineutrinos. The energy of these particles must be approximately equal to the energy of quanta of today's cosmic electromagnetic radiation background, and their concentration is also little different from that of relict quanta.

Experimental detection of relict neutrinos would be an extremely

interesting event. Indeed, the Universe becomes practically transparent to neutrinos only a fraction of a second after the Big Bang. Having detected relict neutrinos, we would be able to look directly into the remotest past of the Universe, by extracting information carried by these particles.

Unfortunately, the detection of neutrinos of such low energies as we expect for relict neutrinos is practically unfeasible for the foreseeable future.

In this connection, we want to mention that neutrino astronomy is being born in our time. We are at the threshold of the stage of systematic analysis of the neutrino flux produced by nuclear reactions in the Sun's core. These neutrinos allow us to look directly into the center of the Sun because its entire mass is absolutely transparent to them. The 'X-raying' of the Sun by neutrinos will provide new information on the internal structure of our star. Likewise, astrophysicists will attempt in the future to accomplish 'neutrino-raying' of the entire Universe.

We have thus gone through the evolution of matter and radiation during the first second after the Big Bang. Even though the possibility of calculating the processes occurring during the first second of expansion may seem to resemble science fiction, modern physics does allows it to be done with complete reliability.

The first five minutes

A familiar song from a Soviet film says:

> Five minutes, five minutes,
> If one is thorough about them,
> One can accomplish lots of things
> Even in these five minutes ...

The first five minutes in the life of our Universe. These minutes predetermined its specific features, including those that manifested themselves billions of years later, in our epoch.

The processes that followed the very first moments mentioned above, and that filled the first five minutes with the drama of awesome nuclear forces, determined the significant features of the chemical composition of today's Universe.

Owing to these processes, stars possess sufficient stores of nuclear energy. Hence, the fact that stars shine is also a corollary of the violent processes that ruled supreme in the Universe in the first five minutes of expansion.

Stars and other celestial bodies grew out of the minute admixture of ordinary matter that we chose to 'forget about' when discussing photons and particle–antiparticle pairs in the preceding section.

Let us now return to this tiny admixture of ordinary matter that we find in the 'boiling cauldron' of neutrinos and antineutrinos, electrons and positrons, and light quanta, during a fraction of the first second of expansion. It has been shown that the processes involving this ordinary matter were extremely sensitive to the conditions that had reigned during the first seconds of expansion. These processes dictated the chemical composition of the matter from which galaxies and stars grew in a much later epoch. As a result, the chemical composition of stellar material is an extremely sensitive indicator of the physical conditions that existed at the start of cosmological expansion.

Let us look at processes in which ordinary matter participates. What is the state in which we find it?

First of all, there can be no neutral atoms at a temperature above 10 billion degrees: the matter is completely ionized and forms high-temperature plasma. Furthermore, complex atomic nuclei cannot survive at this temperature either. A complex nucleus would be immediately smashed into pieces by the high-energy particles that surround it. The heavy particles of matter are, therefore, neutrons and protons. These particles undergo bombardment in the 'boiling cauldron' by energetic electrons, positrons, neutrinos, and antineutrinos.

The interaction with these particles forces neutrons and protons to transform rapidly into one another. These reactions create the equilibrium of protons and neutrons. When temperatures are sufficiently high (above 100 million degrees), neutron and proton concentrations are approximately equal.

As the Universe expands and cools down, the fraction of protons increases and that of neutrons decreases. Their concentrations become unequal because the neutron mass is greater than the proton mass so that the formation of a proton is energetically more advantageous, that is, the probability for a proton to be created is higher than that for a neutron. If the reactions were sustained for longer than several seconds after the Big Bang, the fraction of neutrons would become negligible several tens of seconds later.

Actually, the reaction rate depends critically on temperature. As the temperature drops, the rate of these reactions decreases and they practically stop several seconds after the expansion begins. The relative abundance of neutrons 'freezes' at about 15 per cent of all heavy particles.

After the temperature drops to a billion degrees, the simplest complex nuclei can be formed. The energy of quanta and other particles is now insufficient for decomposing a complex nucleus. All available neutrons are captured by protons, which form deuterons; reactions involving deuterons later produce nuclei of helium atoms. Very small amounts of the helium-3 isotope, deuterium, and of lithium nuclei are also produced.

Practically no nuclei of greater complexity are formed under these conditions. The point is that such elements may be created in appreciable amounts only in collisions of nuclei and particles that have already been produced. This means that more complex nuclei may form when helium-4 nuclei collide with neutrons, protons, or other nuclei of helium-4. However, these collisions cannot produce more complex nuclei with atomic masses of 5 or 8 since no such stable nuclei exist!

As a result of these factors, the synthesis of elements at the start of expansion is limited only to the lightest elements and is completed some 300 seconds after the Big Bang, when the temperature drops to below one billion degrees and the energy of particles becomes too low to produce nuclear energies. Reactions leading to the formation of helium are similar to those occurring when a hydrogen bomb explodes. Elements heavier than helium are synthesized in stars in our epoch. Matter resides in stars for a long time, so that not even the fastest reactions run through. The synthesis of elements heavier than iron occurs in violent, explosive processes (in flares of supernovas). The gas subjected to nucleosynthesis processes in stars is then partly ejected into surrounding space in the slow flow of gas from the star surface and in explosions. This is the gas from which stars of subsequent generations and other celestial bodies will be formed later.

Let us return to the synthesis of light elements at the start of cosmological expansion. Since almost all neutrons went into the creation of helium atoms, it is not difficult to calculate how much helium is formed. Each neutron is included into a helium-4 nucleus with a proton, so that the weight fraction of helium equals twice the neutron concentration, that is, 30 per cent.

We thus find that roughly five minutes after the Big Bang, the matter consists of helium nuclei (30 per cent) and protons, that is, hydrogen nuclei (70 per cent). This chemical composition remains unchanged until galaxies and stars begin to form and nucleosynthesis starts deep inside stars.

Is the conclusion on this chemical composition of pre-stellar matter supported by observations?

How much helium is there in the Universe?

There is very little helium on the Earth. Its scarcity is caused by specific properties of this element and by the conditions under which the Earth was formed and then evolved. Being a very light and inert gas, helium escaped from the Earth's body. However, astronomers find it everywhere even though it is not easy to observe by conventional spectral analysis.

Helium is found in hot stars, in large gaseous nebulae that surround young hot stars, in the outer envelope of the Sun, and in cosmic rays, that is, in the flow of high-energy particles that arrive at the Earth from space. Helium has also been detected in the most remote objects that we observe in the Universe: the quasars.

It is significant that wherever helium is discovered, it almost always makes up about 30 per cent of the total mass, the rest being hydrogen. The admixture of other elements is quite small. Their fraction varies among different objects while the fraction of helium remains impressively constant.

The reader will remember that the hot Universe model did predict these 30 per cent of helium in the primordial matter. If most of the helium had been synthesized in the first minutes of expansion while heavier elements were synthesized much later in stars, the helium abundance would be 30 per cent everywhere and that of other elements would vary, depending on local conditions of synthesis in stars and on subsequent ejection from stars into the surrounding space.

Helium is also synthesized in the course of nuclear reactions, but the fraction of helium produced in this way is small in comparison with the fraction created at the beginning of the cosmological expansion.

Would it be reasonable to hypothesize that the observed 30 per cent helium abundance can be explained entirely by helium production in stars?

No, this hypothesis is definitely unacceptable. First, when helium is produced in stars, large amounts of energy are released, making stars shine with high intensity. If the observed helium were synthesized in stars, the high-temperature light they would have emitted would be observable in the current Universe; this light is not observed.

It can be added that a study of the oldest stars, which definitely were formed from the primordial matter, reveals that they also contain 30 per cent of helium. Hence, practically all helium of the Universe had been synthesized during the first moments of the Big Bang.

Chemical analysis of the matter in today's Universe thus provides a direct confirmation of our understanding and interpretation of the processes that took place during the first seconds and minutes after the Universe began to expand.

Three hundred thousand years of the photon plasma era and the current epoch

One more type of process occurred in the expanding plasma during the first 100 seconds. The point is that after 10 seconds of expansion from the singular state, the temperature in the Universe dropped to several billion degrees. Before that, the Universe contained numerous electrons and positrons created in collisions of high-energy particles. Now the collision energies become insufficient to produce electron–positron pairs. Electrons collide with positrons and are annihilated, turning into photons. The entire energy contained in electrons and positrons converts into photons of cosmic radiation background.

Minutes pass, expansion continues, and the temperature keeps decreasing. The annihilation of electrons and positrons is completed, nuclear reactions in the matter die out.

These were the last active processes in the hot early Universe. It now became too cold (colder than a billion degrees!) to sustain violent processes.

The great fireworks in the life of the young Universe were over and a long quiet period settled in. It lasted about 300 thousand years.

You will remember that throughout this period, the expanding plasma is still very hot and totally ionized. It is opaque to the cosmic radiation background whose mass is greater than that of the opaque plasma. This mixture of plasma and light undergoes slight oscillations that can be called 'photonic sound' because the pressure of light is the force that causes them.

This is about all that was of interest in that 'quiet' era.

The quietness lasted until the time when the temperature dropped to about four thousand degrees. This temperature was sufficiently low for the ionized plasma to begin transforming into neutral gas. This event may not seem so important but it was a turning point in the subsequent evolution of the Universe.

Until this moment, the ionized gas was completely opaque to the radiation background. After the gas (mostly hydrogen) became neutral, it also became transparent to the majority of photons of the cosmic radiation background. At this moment, relict radiation separated from matter. The entire Universe was transparent to it. Photons propagated through matter with practically no absorption, while the matter was expanding and becoming more rarefied and cool.

'Why is it so important?' the reader may ask. It is crucial because now celestial bodies can grow in the cooled neutral gas.

The photonic plasma era is replaced with the era of structure formation in the Universe.

It can be said that the current epoch in the history of the Universe began with the growth of huge blobs of matter in the primordial, almost homogeneous, medium; galaxies and their clusters later evolved from these blobs. The blobs coalesced owing to gravitational forces, so that the process as a whole is called the 'gravitational instability'.

Newton is known to have mused on the possibility of uniform matter coalescing into a blob or several blobs in response to gravitational attraction of particles of matter. He wrote:

If the matter of our sun and planets and all the matter of the universe were evenly scattered throughout all the heavens, and every particle had an innate gravity toward all the rest, and the whole space throughout which this matter was scattered was but finite, the matter on the outside of this space would, by its gravity, tend toward all the matter on the inside and, by consequence, fall down into the middle of the whole space and there compose one great spherical mass. But if the matter was evenly disposed throughout an infinite space, it could never convene into one mass; but some of it would convene into one mass and some into another, so as to make an infinite number of great masses scattered at great distances from one to another throughout all that infinite space. And thus might the sun and fixed stars be formed.

Hence, uniform matter tends to separate into blobs, being driven to separation by gravitation. This has not happened, however. In fact, if this process did occur at the birth of the Universe, neither galaxies nor stars could grow later. Indeed, the matter was tremendously dense at that time. Blobs would have had even higher density. We do not observe it in the Universe; at any rate, we observe very little of it. The average density of galaxies is quite modest. Hence, they coalesced in a comparatively recent epoch, when the expanding matter of the Universe grew sufficiently tenuous. The gravitational instability manifested itself only at this stage. Something prevented this mechanism from 'doing the trick'. This 'something' was the pressure of the cosmic radiation background.

The pressure of relict protons is gigantic. If a blob of increased density grew somewhere in the plasma, the gravitational forces would certainly tend to enhance the buildup of the blob, in accordance with Newton's description. However, powerful forces of counterpressure due to photons, to which the plasma was opaque, acted against gravitational force. Counterpressure tended to spread out the blob and suppressed the manifestations of the gravitational instability.

These manifestations became possible only after the hot plasma turned into neutral gas. Now the gas has become transparent for the relict radiation. Photons escape from the blob unobstructed, so that the gas blob formed in the process of gravitational contraction meets no resistance of pressure due to photons. Only the pressure of the gas can produce counterpressure. However, this counterpressure is much weaker than that of photons, and the forces due to gas pressure cannot overcome gravitation if the blob is sufficiently large. The gravitational instability wins.

Before turning to specific manifestations of the gravitational instability, we first have to consider another puzzle that researchers encountered.

4: Neutrino Universe or '...ino' Universe?

Neutrinos

The neutrino! This particle had brought surprises for physicists before, and was expected to bring more. But nobody anticipated what actually happened in 1980. The picture as scientists saw it in their mind's eye seemed fantastic, to say the least.

Let us approach it in a systematic fashion.

The first surprise was how the Swiss physicist Wolfgang Pauli invented this particle in 1930. 'Invent' is the word that Bruno Pontecorvo, one of the creators of modern neutrino physics, uses when describing the theoretical prediction of the neutrino.

He writes, recalling that period:

It is difficult to find a different situation in which the term 'intuition' would better reflect the nature of scientific achievement than in the case of the prediction of the neutrino by W. Pauli.

First of all, only two 'elementary' particles were known to physics fifty years ago, namely, the electron and the proton, and even the idea that a new particle must be introduced for better understanding of nature was in itself revolutionary . . .

Second, the putative particle, the neutrino, was to possess absolutely exotic properties, and in particular, tremendously high penetrating ability.

Pauli 'invented' this puzzling particle in order to explain the fate of part of the energy released in the radioactive decay of nuclei accompanied with the emission of electrons. This is the so-called beta-decay.

The thing is that when one measures the energy of the products of beta-decay of a radioactive element, for example, of the decay of tritium into helium, one discovers that the sum total of the energies of all recorded particles after the decay is different in different decay events. The law of energy conservation is patently violated, since part of the energy always disappears.

Even such giants of physics as, for example, Niels Bohr began to accept the suspicion that energy conservation is indeed violated in these processes. This is the situation in which Pauli came up with his 'invention'. He assumed that the law of energy conservation is not violated but that the decay produces, in addition to the particles that instruments record, a particle of a new species. These hypothetical particles interact very weakly with ordinary matter and fly away from the laboratory without being stopped by the detectors. The escaping particles carry away the energy that is unaccounted for, making it look as if energy vanishes. The mysterious particles were called *neutrinos*.

More than half a century has elapsed since then; as we have mentioned, the neutrino was to present physicists with quite a few puzzles. Thus it was found that neutrinos interact with matter not just weakly but fantastically weakly. They travel freely across, say, the Earth, or Sun, stars, or any object in the Universe as if it were empty space, as light travels through glass in a window.

Obviously, this property makes detecting such particles an extremely difficult task. They were discovered directly, via the nuclear transformations that they cause, only in 1953–6.

Further analysis revealed that there are at least three species of neutrino (in this presentation, we will not distinguish between the neutrino and its antiparticle, the antineutrino; we employ the common term *neutrino*): electron, muon, and tau neutrino. Each species participates only in reactions specific for it.

We will not go though all the amazing features of the neutrino here. It need only be pointed out that these unusual features were so mystifying that physicists, on the one hand, helplessly recognized their inability to fathom the reasons for neutrino characteristics but, on the other hand, anticipated with very nearly religious certainty (rather, their scientific intuition supported this attitude) that a particle so strange must play a very special role in the Universe. Here is what distinguished physicists said about it two decades ago.

John Archibald Wheeler, the former President of the American Physical Society, remarked that physicists knew at that time of no explanation at all why neutrino interactions have to be so weak in comparison with the electromagnetic interactions, or why they are so strong in comparison with the gravitational interaction.

It is remarkable that Wheeler placed this one sentence as a separate chapter [*sic*] in his book *Neutrinos, Gravitation and Geometry*. Note, for comparison, that, say, the first chapter in this book contains a hundred-odd pages with very complicated formulas.

Markov, who greatly contributed to the progress of neutrino physics, wrote:

It is difficult for a contemporary to anticipate the role of the neutrino in future physics. However, the properties of this particle are so elementary and so unusual that it is only logical to believe that nature created neutrinos to achieve profound, not yet comprehensible for us 'purposes'.

We will see later what these 'purposes' are.

The recent discoveries that we will discuss make us look with even greater attention at the neutrino and re-evaluate the combination of the three grand essentials: gravitation, neutrino, and the Universe.

While gravitation is the main force that drives the motion of matter in the Universe, the neutrinos, or some other particles that resemble the neutrino in the property of 'virtually not interacting' with the rest of matter, now appear to be the most important particles in the Universe. Such putative particles abound in modern theories. Even though they have not yet been detected by experimentalists, some of them look like very realistic candidates. Among such particles are the gravitino, photino, and a number of others, all invented by theorists. We refer to all of them as '. . .inos'. It is about these '. . .inos' that one has to think first of all when trying to understand what the Universe is.

The properties of the Universe

The preceding chapters have already introduced the reader to some important properties of the macroscopic world around us; these properties were discovered and then checked and verified by science. Let us recall some of these undoubtedly true facts that are necessary for further discussion.

We remember, first of all, that the expansion of the Universe began from a superdense state when the matter was extremely hot. The cold cosmic microwave radiation background is a vestige of that epoch. It has also been reliably established that no appreciable inhomogeneities exist in the distribution of matter on the scale of billions of light years: supersuperclusters of galaxies do not exist. This means that a very-large-scale Universe does not consist of individual structural units. This fact has been established especially well by observing the cosmic microwave background: if inhomogeneities of the order of a billion light years or greater existed, the cosmic microwave background would arrive at us with different intensities from different directions. This would happen because enhanced density leads to an enhanced gravitational field. The microwave background photons leaving this gravitational field lose additional energy, that is, become 'redder', so that the intensity coming from these directions must have slightly lower intensity. Since no such variations in the intensity of the cosmic background are observed, the hierarchical pyramid of structural units in the Universe does not stretch to infinity. In other words, the Universe is homogeneous on a very large scale, beginning with regions of about a hundred million light years.

Recall also that observations revealed important features of the largest structural units in the Universe, that is, of superclusters of galaxies. It was found that galaxies and their clusters are located in thin layers that form the walls of cells whose internal regions are virtually empty. The distribution of galaxies in the Universe can be said to resemble 'honeycombs'. The density of galaxies is especially high in the edges of honeycomb 'cells'.

Some important facts about the structure and evolution of the Universe are thus very reliably established: the expansion of the Universe, its initial hot state, and the current cellular structure.

Unsolved problems

Among such problems, we will begin with the identification of the mechanisms of inception and growth of structure in the Universe.

How has the present structure of the Universe arisen, when, and why? Why are the largest structural units of the Universe, that is, large clusters and superclusters of galaxies, of the scale and shape that we observe them to possess, not different? Astrophysics theorists, in collaboration with observers, have tried to answer these questions for the past 15 years, but even very recently it was not yet possible to say that the main stages of the process of formation of galaxies and their clusters have been identified and understood.

Indeed, something very important remained unknown. A suspicion that our knowledge of the Universe has an essential gap has been felt since quite long ago, from the period when astrophysicists came across the so-called dark-matter puzzle that was outlined in one of the preceding chapters.

You will remember that this problem had been clearly formulated at the beginning of the 1970s. It is as follows. The motion of galaxies in their clusters is such that one has to assume that there is some invisible mass in the space between galaxies. This mass affects the moving objects by its gravitation but otherwise does not manifest itself in any way. In all likelihood, similar dark matter surrounds large galaxies; this is inferred from the motion of dwarf galaxies and other objects around them. This invisible mass is called the dark matter, or hidden mass, and practically nothing was known about its nature. Observations indicated that the regions of concentration of galaxies contain about 20 times as much dark matter as they have visible matter in galaxies themselves. The mass of all galaxies in a typical cluster is about 3×10^{13} solar masses but the mass of invisible matter is found to be about 10^{15} solar masses. True, some specialists were of the opinion that the observations revealing the manifestations of the dark matter are not sufficiently reliable; arguments about this raged with variable intensity until very recently.

Neutrinos in the Universe

Now we return to the protagonist of our story, the neutrino. Something must be added to what we presented at the beginning of this chapter. Even very recently physicists did not question the belief that neutrinos have zero 'rest' mass and move at the speed of light.

Processes involving neutrinos and likely to play important parts in astrophysics were carefully studied for some time.

Thus it was established that the number of neutrinos in the vast Universe is almost as high as that of relict electromagnetic quanta (photons). As we saw in the preceding chapter, both photons and neutrinos have survived in the Universe since that initial stage of expansion when the hot dense matter had an extremely high temperature and was opaque not only to photons but to neutrinos as well. This was the time of fast reactions transforming neutrinos, electrons, electromagnetic quanta and other elementary particles into one another. These processes allow reliable calculations to be carried out in the framework of modern physics. The results of such calculations show that several tens of seconds after the Universe began to expand, the number of neutrinos (together with antineutrinos) in one unit volume was approximately one third of the number of photons.

This ratio of relict photons and relict neutrinos has remained almost unchanged in the course of the subsequent expansion of the Universe, until the present day. So far, we have been unable to detect relict neutrinos in any direct way since their energy is exceedingly low: if neutrino mass is zero, its energy is about 5×10^{-4} electronvolt (eV). Nevertheless, astrophysicists are capable of predicting their abundance. We have already mentioned that each cubic centimeter contains about 500 relict photons. The number of relict neutrinos is about a third of that, that is, about 150 particles per cubic centimeter.

The reader remembers that each photon of the cosmic microwave background has a certain energy, and the mass corresponding to it is 10^{-36} g, so that the mass density of the cosmic background radiation is about 5×10^{-34} g/cm^3. This is approximately 2000 times less than the mean density of normal matter in the Universe.

We can conclude now that the mass density of the cosmic microwave radiation background is negligibly small. The same could be said about neutrinos: the mean mass density (determined by the energy of these particles) is still lower than that of the electromagnetic radiation: it is around 1.5×10^{-34} g/cm^3. Therefore, the role of cosmic background neutrinos in the contemporary Universe can be safely ignored: not only is their total mass negligible, but they practically do not interact with the rest of the matter in the Universe.

This, at any rate, was the opinion of most of the specialists about the role played by neutrinos in today's Universe – until the spring of 1980.

Neutrino experiment

In the spring of 1980, a group of physicists at the Institute of Theoretical and Experimental Physics (ITEP) of the USSR Academy of Sciences, headed by Lyubimov and Tretyakov, published the results of many years of experimental work; the results pointed to nonzero mass of the electron neutrino. (Recall that for the sake of brevity, we speak only about electron neutrinos. In fact, two more species of neutrino are known: the muon and the tau neutrinos.) The probable mass of the electron neutrino (their 'rest mass') as obtained in this experiment was about 6×10^{-32} g, or, in energy units, 35 eV. This means, among other things, that contrary to earlier belief, neutrinos need not move at the velocity of light but may move at any velocity below that of light, or be at rest.

I want to emphasize the staggering complexity of the experiment aimed at measuring the neutrino mass, and also the fact that the experimenters themselves did not insist that their result is the ultimate true neutrino mass. This figure would be checked and rechecked in subsequent experiments. If, however, this result could be confirmed, its corollaries will be very serious, especially for astronomy. It is mostly for this reason that theorists did not bother to wait for the ultimate confirmation of the results on the evaluation of the neutrino mass and began to analyze what has to be changed in our picture of the Universe if the neutrino is massive. Other experiments have been reported recently that point to nonzero neutrino mass, and not only for the electron neutrino.

Actually, possible corollaries for astrophysics implied by the hypothesis of nonzero neutrino mass had been discussed long before these experiments. As early as 1966, the Soviet physicists Gershtein and Zel'dovich analyzed the effects of considerable neutrino mass on the expansion of the entire Universe. Two Hungarian theorists, Marx and Szalay, also studied cosmological consequences of the assumption of nonzero neutrino mass.

However, these were all only tentative trials, an analysis of various possibilities. The situation changed drastically after the direct experiment of the Soviet physicists.

Being prodded and stimulated by experimentalists' indication, theorists started the veritably massive assault on the problem.

The neutrino Universe

From the data obtained by the ITEP physicists, the neutrino is 20 000 times lighter than the electron and 40 million times lighter than the proton. Why then do theorists expect this lightest particle that does not interact with anything to be a decisive component of the Universe?

The answer is very simple: relict neutrinos are extremely numerous. Generally, their number in one cubic centimeter is greater than that of protons, on average, by a factor of one billion; hence, even though each neutrino has a very tiny mass, the total mass of neutrinos constitutes the predominant component of matter in the Universe. It is not difficult to calculate that if the mass of the electron neutrino is 6×10^{-32} g, then their mean density (ignoring other sorts of neutrino) is about 10^{-29} g/cm^3, which is greater by a factor of 10–30 than the density of the entire 'non-neutrino' matter. As a result, it is the gravitation of the neutrinos that claims the role of the dominant force that dictates the laws of expansion in today's Universe. The mass, and hence, the gravitational effect, of the ordinary matter comes to an only 3–10 per cent 'admixture' to the main mass of the Universe, that is, its neutrino mass. One is then permitted to say that the Universe mainly consists of neutrinos, that we live in a neutrino Universe. It was this conclusion that we meant when hinting at the beginning of the chapter at a fantastic picture that the scientists discovered.

Another consequence of the above conclusion can be pointed out.

The question about whether the expansion will be eternal or finite is a most important aspect of the evolution of the Universe. As we know, the answer depends on the mean density of matter in the Universe: if this density exceeds the critical value, the gravitation of matter will slow down and, after some time, stop the expansion of the Universe and force galaxies to begin moving closer again; the expansion of the Universe will be replaced by its contraction. If the density of matter is less than the critical value, the gravitational pull of the matter will be insufficient to stop the expansion and the Universe will remain forever expanding.

From current data, the critical density is, as we have mentioned, 10^{-29} g/cm^3. Until very recently, it was believed that the main fraction of the density of the Universe was contributed by the ordinary matter whose density is approximately 3×10^{-31} g/cm^3. This estimate signified that the density is below the critical level and the Universe is doomed to eternal expansion. Now we discover serious arguments in favor of estimating the density of relict electron neutrinos (without other types of neutrino) as approximately equal to the critical value of 10^{-29} g/cm^3. We should not forget that in addition to relict electron neutrinos, there also exist muon and tau neutrinos. Nothing is known about their mass from direct experiments but theory and indirect experiments imply that if the electron neutrino mass is nonzero, the masses of the other sorts of neutrino are unlikely to be zero. Moreover, the muon and tau neutrino masses are probably not lower than the electron neutrino mass. If these conjectures are accepted, the mean density of matter in the Universe is greater than the critical level. This means that in the very remote future, many billions of years after our time, expansion of the Universe will have changed to contraction. Note that the reason for this 'strongest' inference is the 'weakest' of all particles, the neutrino.

The origin of galaxies

Let us return to the origin of the structure of the Universe. At the initial phase of the expansion, the matter was nearly uniform expanding hot plasma. What made this uniform plasma separate into the blobs that later evolved into celestial bodies and their systems? What led to the nucleation of clusters of galaxies?

According to the opinion of most specialists, this process was caused by gravitational instability: random small-amplitude initial density increases attracted surrounding matter by their gravitation and were thereby enhanced: they became denser and larger. Under certain conditions, these blobs of matter could grow into the very large blobs that later gave rise to clusters of galaxies. The foundations of the theory describing this process were formulated by the Soviet physicist Lifshitz as early as 1946.

We can assume now that the gravitation of neutrinos is the most important factor in the Universe; namely, this gravitation must be treated as the uppermost factor in the analysis of the growth of inhomogeneities of matter driven by the gravitational instability.

The general picture of the growth of inhomogeneities appears to be as follows. At the very first moments after the expansion began, there were random, very tiny inhomogeneities in the distribution of density in space. We know that just one second after this instant of time, the density of matter fell enough to become freely traversable by any type of neutrino. At this period, neutrinos still possess very high energies and move at velocities very close to that of light. Obviously, they level out the small inhomogeneities and the neutrino distribution becomes more uniform. However, this occurs only over small space scales in regions whose linear dimensions are relatively small in comparison with those of neutrino blobs.

Indeed, neutrinos from relatively small blobs manage to escape and mix with other neutrinos sufficiently fast, thereby averaging and smoothing all inhomogeneities. The later the moment of time, the larger the inhomogeneities that manage to 'dissolve'. This process lasts until neutrinos begin moving at velocities considerably lower than that of light, because the expansion of the Universe decreases their energies. Calculations show that about 300 years after expansion begins, the velocity of neutrinos decreases so much that they fail to leave very large blobs. Blobs whose density was originally only slightly above the average value can be enhanced by gravitation, and grow denser and bigger, until finally the medium decays into separate contracting clouds of neutrinos.

We can calculate the mass of such neutrino clouds. Since the smoothing of density and the motion of neutrinos at nearly the velocity of light lasted only for the first 300 years, we conclude that smoothing occurred in regions whose size did not exceed 300 light years. On a large scale, in still greater neutrino blobs, an increased neutrino density was retained because neutrinos did not have enough time to leave them. Then the velocity of motion of the neutrinos sharply decreased, their mutual gravitational attraction led to an additional increase in density, and these new blobs gave rise to neutrino clouds. Consequently, the mass of these clouds is determined by the number of neutrinos contained within a sphere with a radius of 300 light years, 300 years after the Universe began to expand.

Calculations show that the typical mass of such a neutrino cloud is expressed only in terms of the fundamental constants of nature: Planck's constant h, the velocity of light c, the gravitational constant G, and neutrino mass m. The first three constants are known and if we

assume that the neutrino mass is indeed $35 \text{ eV} = 6 \times 10^{-32}$ g, we find that the mass of a typical neutrino cloud is about 10^{15} solar masses.

This is the situation with the mass of neutrino clouds. What do we know about their shape? Fifteen years ago, Zel'dovich was able to show that clouds created in processes of this type must be quite flattened, resembling pancakes in their shape. A combination of numerous 'pancakes' randomly distributed in space produces a pattern of an invisible neutrino honeycomb.

We thus conclude that a cellular structure of invisible neutrino clouds must have formed by now in space; the same applies to possible clouds of other '...inos'. And what about ordinary matter? Does it gather to form a spatial structure?

At the beginning of the expansion, the ordinary matter (all matter in the Universe except neutrinos) was also distributed in the Universe almost uniformly. As we know (or rather, as we have grounds to believe), the mass of this matter was much smaller than the total mass of neutrinos, and at the initial stage of expansion of the Universe this matter formed hot plasma.

We have seen in the preceding chapter, however, that about three hundred thousand to a million years after the Big Bang, the ordinary matter cools down so much that the plasma turns into neutral gas whose pressure drops sharply. Then the cold neutral gas begins to get denser in the gravitational field of the neutrino clouds as they appear and concentrates in their central regions. These denser blobs of neutral gas gradually give rise to clusters of galaxies, galaxies, and stars. As the mass of the ordinary matter is one-thirtieth of the neutrino mass, a large cluster of galaxies with a mass of about 3×10^{13} solar masses is formed within an invisible neutrino 'pancake' of 10^{15} solar masses.

The data obtained from the observational astronomy on the mass and shape of large galactic clusters definitely point to the presence of clouds of invisible particles, the '...inos'. In fact, the assumption that these are clouds of conventional neutrinos does entail certain difficulties; in this case, other '...inos' are responsible.

Realities and science fiction

The vast sea of neutrinos, assembled into clouds in which they move at a velocity of the order of 1000 kilometers per second, represents, presumably, the very 'something' that had earlier been overlooked in the study of the Universe and without which it is impossible to interpret many of its important features.

Astrophysics theorists say that many incomprehensible facts found proper places in the general picture after it became justifiable to regard the neutrino, or other '. . .inos', as massive particles. The Soviet astrophysicist Doroshkevich said in 1980 in a periphrasis of a well-known adage that 'if the neutrino mass were found to be zero, we would have to invent some other particle of nonzero mass that would very weakly interact with all other particles'. It was such particles that we referred to as '. . .inos'.

In fact, a 'replacement' particle mentioned in Doroshkevich's half-joking remark has already appeared on the list of hypotheses of modern physics. As an example, we can again mention the photino, which is a photon-like particle but with nonzero mass, and the gravitino, which is similar to the graviton but is also massive. So if the neutrino mass is proved to be negligibly small and the mass of the world is not determined by neutrinos, then the Universe is an even more curious combination, dominated by, say, photinos or gravitinos, or by some other '. . .inos'.

We are as yet quite far from reaching the truth in this, and much of what we have outlined lies at the forefront of today's science. Correspondingly, I have tried to differentiate between reliably established facts and problems that are being considered now.

The famous British physics theorist Stephen Hawking invited scientists to Cambridge, in the summer of 1982, to a small-scale international workshop to discuss processes that occurred in the Universe before the first second elapsed after the expansion had begun. We will discuss these processes later. Once, late at night, Professor Markov and I were walking after a day of intensive and interesting work along the narrow streets of this old, and possibly most famous, center of science in the world. Unexpectedly, our discussion turned to the extent to which the picture of the Universe as we compose it today has become fantastic, diversified, and amazing. It is so much richer than the mechanistic picture of the motion of indivisible spheres as it seemed to the great Newton who had been pursuing his research in that same town several centuries ago.

I reminded Professor Markov of his prophesy on the role of neutrinos in the Universe (it was quoted at the beginning of this chapter) and remarked that things that we, specialists, were discussing now at our workshop were incomparably more fantastic than anything encountered in science fiction. Markov replied that there was no such thing as serious science-fiction literature. Any serious, or true, literature always centers on people, on their souls. A writer may employ situations from a 'science dreamworld', creating what is known as science fiction (it may be good, or it may be bad). But any attempt at being 'scientific' in science fiction is dilettantism, and literature of this sort ceases to be fiction, without even gaining a semblance of science. In contrast to this, true science is always fantastic! To comprehend it, and especially to develop it further, one needs an uncommonly powerful imagination which nevertheless operates with rigorous formulas supported by a solid foundation of knowledge. 'To be a scientist is not easy, although very interesting', remarked Professor Markov.

As for science fiction, it became clear that the professor is not only an avid reader of this literature but writes in this genre himself. Back in Moscow, I was able to read a short science fiction novel written by M. Markov.

Coming back from fantasy to realities, let us summarize the results of our journey into the first seconds in the life of the expanding Universe. We witnessed violent processes in the early hot Universe, the real fireworks that gave rise to the birth of numerous worlds and to the Universe as we know it.

Today we live in the Universe with well-developed structure and with systems of worlds. A directed process of converting hydrogen into helium and heavier elements is going on in stars. The stores of nuclear fuel are enormously rich, sufficient to sustain the reactions for tens of billions of years. What comes next? Stars cannot be an eternal attribute of the Universe: they burn out and die. One of the outstanding cosmologists, G. Lemaître wrote: 'The evolution of the world can be compared to a display of fireworks that has just ended: some few red wisps, ashes, and smoke. Standing on a cooled cinder, we see the slow fading of the suns, and we try to recall the vanished brilliance of the origin of the worlds.'

Does this mean that in the future the Universe will resemble smouldering ruins left after a great fire?

Definitely not! We will discuss the future of the Universe later in the book. But first we have to approach again the singularity from which the Universe began to expand. This time we will come much closer to the singularity, and will need wings of scientific imagination (not science fiction!), the very imagination that Professor Markov was emphasizing.

5: At the frontiers of the known

Why is the Universe the way that we see it?

What was going on in the Universe very close to the singularity, at temperatures much higher than the 10^{13} kelvins at which we had stopped in the preceding chapters?

We are already familiar with the general method used to probe the processes that occurred at the very beginning of the cosmological expansion. Namely, one looks for the 'traces' left by these processes. We have already mentioned that one bright 'imprint' of the processes taking place within the first several seconds of the Big Bang is the chemical composition of the pre-stellar matter: the 30 per cent of helium synthesized in that remote era. Now we want to look for, if possible, just as explicit 'traces' of still more 'ancient' processes.

It was established that the fundamental properties of the Universe are precisely the 'traces' we want to find. Let us begin by listing them, and then try to identify the processes responsible for these properties; we will discuss how current science hopes to explain these mystifying properties of the Universe.

The first enigmatic property is the enormous number of photons of the cosmic radiation background as compared with the number of heavy particles. The reader will recall that their ratio is a billion to one. What caused this huge difference?

The second puzzle: why is the Universe so very homogeneous on a large scale? As we know, the homogeneity is established reliably by the observation of the cosmic microwave background coming to us from outer space at direction-independent intensity. This means that at the moment in the past when the plasma turned into neutral gas and thus became transparent, and the relict photons that we observe today were emitted, very distant points of space had the same temperature. For that early period, each such point lay outside the apparent horizon drawn around the other points. These points could not, therefore, be causally connected: they could not exchange signals. How, then, could these temperatures be identical if one point could not even find out what the temperature was at the other points. This problem is known as the 'horizon problem'.

The third puzzle: why is it that today, 10–20 billion years after the Big Bang, the density of matter in the Universe is fairly close to the critical value and the geometrical properties of space are very close to those of flat space? Note that if the matter density at some moment deviates from the critical density, the difference can only increase with time. Indeed, if density equals the critical value, it means that the rate of expansion is exactly balanced by gravitational forces. If this balance is slightly violated, for instance, in favor of gravitation, the slowing down of the expansion will, after some time, violate the balance. Therefore, if today's density of matter differs from the critical value by less than an order of magnitude, the balance of gravitation and expansion rate in the past had to be tuned with fantastic precision. It can be calculated that one second after the expansion began, the balance could not be violated by more than one-ten-thousandth of one-billionth of one per cent! What factors resulted in this precision?

Another puzzle: despite the amazing homogeneity of the Universe on a very large scale, deviations from uniform distribution on a smaller scale did occur in the past. What caused these small primordial fluctuations that later evolved into galaxies and their systems? This is the problem of primordial fluctuations, not just any fluctuations but those that resulted in the formation of individual worlds in the epoch that is not very remote from ours.

A key to solving these problems was found in the advances in elementary particle physics.

Let us see how this key unlocks the chest with nature's innermost secrets.

We know about four types of physical interactions: the strong (or nuclear), electromagnetic, weak (dictating, e.g. radioactive decay), and gravitational. According to current concepts, these types of interaction manifest themselves as distinct interactions only at relatively low energies, and merge into a unified interaction at high energies. Thus at energies of the order of 10^2 gigaelectronvolts (GeV), which corresponds to a temperature of 10^{15} K, the electromagnetic and weak interactions merge. At energies of about 10^{14} GeV, or a temperature about 10^{27} K, the so-called 'grand unification' of the strong, weak, and electromagnetic interactions takes place. And finally, the gravitational interaction is expected to merge with those three at energies about 10^{19} GeV, or temperatures about 10^{32} K ('superunification').

Let us put aside the possibility of the final unification of all forces with gravitation and look at the cosmological consequence of the 'grand unification' theory.

We begin with the first of the problems listed above. The reader may have been slightly surprised by this problem being classified among puzzles. What is so puzzling in a billion relict photons per heavy particle?

The extraordinariness of this becomes apparent if we travel back into the past to temperatures of the order of 10^{13} K, when, as we already know, an enormous number of particle–antiparticle pairs were constantly created and immediately annihilated. Among these particles there were electrons and positrons, as well as protons and antiprotons, neutrons and antineutrons. The number of particles of each species was then approximately equal to the number of relict photons. The 'boiling cauldron' that we mentioned contained roughly equal numbers of all species of particles and their antiparticle counterparts.

If the numbers of heavy particles and their antiparticles (they are known as *baryons*) were exactly equal within each species, they would all have been annihilated during the expansion phase and transformed into relict photons and neutrinos; nothing would survive in the Universe except for the radiation and neutrino backgrounds! There would be no matter from which stars and planets and then finally people could evolve.

For some reason, however, the numbers of particles and antiparticles were not exactly equal, although the difference was quite small. There was one 'superfluous' heavy particle per billion particle–antiparticle pairs! The billion pairs were annihilated as the temperature decreased but this 'extra' particle survived. The world that surrounds us today – the world of stars, planets, and gas – grew out of these surviving particles.

We have come again to a rather strange situation: a billion pairs and one extra particle. Where did it come from and why is there one such particle per billion?

This is the problem at the top of the list. It had been assumed until recently that if an 'extra' particle did not exist at the very beginning, there could be no way for this particle to appear in any reaction. The assumption was that the 'baryon charge' must be conserved (the baryon charge is the difference between the numbers of heavy particles and antiparticles). Weinberg and Sakharov had already pointed to the possibility of violation of this dogma in the 1960s. Later the theory of 'grand unification' indeed showed that some reactions do violate the law of baryon charge conservation. However, these reactions involve superheavy particles: the so-called superheavy Higgs and gauge particles. These particles can be created only at very high energies, so that the reactions that do not conserve the baryon charge are allowed to take place only at such high energies. Here we will describe just one possible mechanism for producing an excess of particles in the Universe, the one that was the first, historically, to be suggested. To simplify the presentation and emphasize the main idea, we speak of a single superheavy particle, the superheavy X-boson. In energy units, the mass of this particle equals the energy of 'grand unification' – 10^{14} GeV (10^{14} times the proton mass) – that is, X-bosons can be effectively created at energies corresponding to a temperature of 10^{27} K. The Universe cooled to this temperature 10^{-34} s after the Big Bang. At this (and higher) temperatures, the reactions not conserving the baryon charge were as intensive as all other reactions.

Another important factor is the absence of particle–antiparticle symmetry. This means that reactions with superheavy particles may proceed at rates that in general may differ somewhat from those of reactions with antiparticles.

Now we can explain how one 'extra' particle was created in the expanding hot Universe per billion particle–antiparticle pairs.

At temperatures above 10^{27} K, the Universe contained a superhot mixture of all fundamental particles and of exactly the same numbers of their antiparticles; the mixture is in thermodynamic equilibrium, without any excess of 'extra' particles. If it were not for differences between the properties of particles and antiparticles and for reactions violating the conservation of the baryon charge, all pairs of heavy particles and their antiparticles would have been annihilated in the course of expansion and cooling of the Universe (their numbers would be exactly equal), and by our time the Universe would have neither the protons nor neutrons: they would all transform to light particles. Today's Universe would contain no ordinary matter.

In reality, the following processes occur. When the temperature drops below 10^{27} K, the pace of all processes involving X- and anti-X-bosons becomes slower than the rate of expansion of the Universe. These particles fail to be annihilated or decay, and their concentration is 'frozen'. They start to decay only later, after enough time elapses. It is this process that represents the key to what follows.

Both the X-boson and its antiparticle, the anti-X-boson, can decay in a way violating the baryon charge conservation; it is important that X-particles decay not quite as anti-X-particles do. Calculations show that as a result, a slight excess of particles over antiparticles is produced. So far these calculations are not very accurate but they nevertheless indicate that the number of extra particles is probably about one particle per billion particle–antiparticle pairs. As the Universe expands, particles and antiparticles are annihilated and finally convert into photons that together with the already present photons will create the cosmic radiation background (as we know, the Universe will also have a neutrino background); the baryon excess still survives. This excess is what constitutes the ordinary matter of today's Universe. Clearly, the number of photons will be greater than that of 'extra' particles by a factor of about a billion.

Thus the first problem is solved.

Let us turn now to the other three. According to the 'grand unification' theory, the Universe at 10^{27} K and higher contained also a field (called *scalar field*) that possessed the properties of the vacuum that we discussed in the section 'Gravitation of empty space'. In particular, this field had enormous 'negative pressure', or tension equal to the energy density of the field itself. This field is also known as a 'false vacuum'. In what follows we outline only one of the simplest hypotheses concerning the start of the expansion, that is meant to demonstrate the most important ideas. The 'false vacuum' differs from the true vacuum, among other things, in that the density of the false vacuum is extremely high: about 10^{74} g/cm^3. We know that the vacuum density corresponds to the cosmological constant in Einstein's equations of gravitation. At that epoch this constant (by analogy to the 'false vacuum', it can be called the 'false constant') was also immense.

The temperature of the Universe less than 10^{-34} s after the Big Bang was greater than 10^{27} K. The density of the 'false vacuum' was 10^{74} g/cm^3 but the density of the hot real particles and antiparticles of the ordinary matter was even higher. As a result, the gravitational properties of the 'false vacuum' could not manifest themselves in any way, and the expansion of the Universe followed familiar laws. As the expansion continued, the density of ordinary matter decreased and after 10^{-34} s became equal to the density of the 'false vacuum'. We have already seen in the section 'Gravitation of empty space' the extent to which the gravitational manifestations of the vacuum are unusual. Instead of attraction, its gravitation causes repulsion. This is what took place in the 'epoch of 10^{-34} s'. The gravitational repulsion of the vacuum results in an accelerated expansion of the world. The density of the 'false vacuum' is constant, not decreasing with time, so that the acceleration of expansion is also constant. The rate of expansion (the velocity of recession of any two arbitrary elements of the medium) continuously grows (instead of diminishing with time as it would if the gravitation of the vacuum was not added to that of ordinary matter), so that all distances in the Universe stretch enormously and become astronomically large. This stage of accelerated expansion is known as the 'inflationary Universe'. All distances in the Universe grew in this period from 10^{-34} s to 10^{-32} s by a factor of 10^{50}!

But the state of the 'inflationary' Universe is unstable. The temperature and density of ordinary matter decrease drastically during this phase. The Universe becomes supercooled. The density of ordinary matter drops to values that are absolutely negligible in comparison with that of the 'false vacuum'. The instability means that a phase transition becomes possible from the state of 'false vacuum' with its huge density to a state in which the entire mass density (and the corresponding energy density) of the 'false vacuum' transforms into the mass density of ordinary hot matter while the density of the true vacuum is zero or vanishingly low. This means that the energy that was contained in the 'false vacuum' gives rise to an enormous number of particles and antiparticles of ordinary matter, all having very high energies. The Universe again heats up to a temperature of about 10^{27} K.

We will not go into the details of this transition. Note only that it seems that the reheating of the Universe takes place 10^{-32} s after the expansion began. In a short period between 10^{-34} and 10^{-32} s, the Universe is 'inflated' at an enormous acceleration by the gravitational repulsion of the 'false vacuum'. Without the 'inflationary' stage of expansion, the Universe would expand in this interval only tenfold, while the inflationary stage blows it out during the same time 10^{50}-fold! After this, the expansion obeys the laws of evolution of the hot Universe theory with which the reader is already familiar.

Helium synthesis and other processes described earlier in the book take place at a stage much later than the inflationary Universe phase (compare: 1 s–300 s for helium synthesis and 10^{-34}–10^{-32} s for the inflationary stage).

The theory of the 'inflationary Universe' seems to give the correct description of the very beginning of the evolution of our Universe. On the other hand, concrete figures and a number of details will certainly have to be elaborated upon.

The stage of the 'inflationary' Universe immediately solves the second of the problems listed at the beginning of this section: the horizon problem. Indeed, let us consider two points that, prior to the 'inflation', lie very close to each other within the common apparent horizon for the moment we chose. Signals exchange, temperature equalization, and other processes, are possible between them. The subsequent rapid stretching of distances during 'inflation' separates these points to gigantic distances. In our epoch, the spacings between these points are considerably greater than the distance to the horizon, unless we take into account the inflationary phase. Hence, points that definitely cannot exchange signals after the inflation, had ample opportunities to do so before the inflation.

The transition from the density of the 'false vacuum' to that of ordinary matter at the end of the inflationary stage solves the third problem. The 'antigravitation' of the 'false vacuum' compels the ordinary matter, emerging from it, to expand at a precisely 'balanced' rate. It can be said that the density of the vacuum exactly corresponds to the critical density at that epoch, so that for obvious reasons, the matter density after the phase transition also equals the critical value – and with fantastically high accuracy.

Let us turn now to the fourth problem, namely, that of the generation of small primordial fluctuations of density that had to exist in the medium immediately after the end of the stage of 'inflationary Universe'. The quantum nature of matter is sufficient in itself to produce such inhomogeneities as a result of the processes involved. Such processes always produce small inhomogeneities.

Thus when matter decays, some of its parts decay a little faster and others, a little slower. Likewise, the quantum decay of the 'false vacuum' occurred at some places a little earlier than at other places; as a result, the transition to expansion due to the gravitation of the generated hot matter happened at slightly different moments of time at different places, with ensuing slight inhomogeneities of density. This process is nothing other than primordial acoustic vibrations that later lead, after a long evolution, to the formation of galaxies.

This is how the theory of the 'inflationary' Universe explains the main features of the world that surrounds us.

In fact, this theory gives a number of other, quite interesting predictions.

We have already mentioned that the disappearance of the 'false vacuum' can be compared with a phase transition.

We are familiar with phase transitions, for instance, in the process of solidification of a liquid, when it transforms into a solid, crystalline state. When a liquid crystallizes, numerous crystals with different orientations of the crystallographic axes may nucleate at different points of the liquid. As a result, a number of distinct contiguous regions, or domains, arise in the frozen liquid.

According to the recent results of an analysis of the processes in the early Universe, neighboring domains with different properties form in a similar manner in the 'inflationary' Universe when it undergoes the phase transition. Various exotic particles and structures appear at domain boundaries. For instance, so-called magnetic monopoles may be created there. These particles are carriers of isolated magnetic charge, just as electrons or protons are carriers of isolated electric charge. In contrast to them, the magnetic monopole is expected to be superheavy, of 10^{16}–10^{17} proton masses! Such particles cannot be created in today's Universe: there are no sources of such high energy. So far, magnetic monopoles have not been discovered. They can be expected to abound at domain boundaries. Let us follow the evolution of a domain after its birth in the expanding Universe.

Domains sprang up in the 'epoch of 10^{-34} s' after the expansion began. The size of each domain, correspondingly, is about 10^{-34} of one light second, or about 10^{-24} cm. During the subsequent stage of 'inflation' of the Universe, this domain becomes blown up 10^{50}-fold, that is, grows to 10^{26} cm (note that this is already 10 million light years!).

The stage of inflation ends 10^{-32} s later. After this, the Universe expands in accordance with more familiar laws, being decelerated by the ordinary gravitation. By our epoch, distances grow still more, by a factor of 10^{25}. The domain size is then approximately 10^{51} cm. This distance is truly colossal: about 10^{33} light years. The reader may remember that the diameter of the observable part of the Universe is 'merely' 10^{10} light years in diameter! No signal emitted in the Universe after the stage of 'inflation' travels farther than 10^{10} light years. This distance is the apparent horizon introduced earlier.

Therefore, if domains as corollaries of phase transitions in a remote part of the Universe do exist, they are enormously large. We live in one such domain, somewhere inside it. The walls that separate 'our' domain from others are likely to lie at a distance of about 10^{33} light years! Inside the domain, the distribution of matter on a large scale (large by our standards) is uniform. Monopoles and other types of 'exotica' cling to the walls which form a fence between different worlds.

Is not it true that the world we are now contemplating is a most interesting and amazing place?

Our Universe, which is so homogeneous on a tremendously large scale, becomes inhomogeneous again far beyond the apparent horizon! The Universe we were talking about previously was just 'our' domain.

It is logical to turn at this juncture to what we find in the history of astronomy. All systems of the world invented in different epochs invariably claimed to give the description of the entire world and entire Universe; in fact, they were models of specific astronomical systems. Aristotle's and Ptolemy's systems correctly reflected certain features of the Earth as a celestial body: its spherical shape, the motion of the Moon around the Earth. The rest of the system proved to be erroneous. Copernicus's system was a model of the Solar System. Herschel's Universe is a model of our Galaxy. It is quite possible that the properties of the world of galactic clusters happen to describe 'only' our domain.

The properties of the matter surrounding us are truly inexhaustible, and the power of the human brain trying to comprehend these properties is equally unlimited.

All aspects touched upon in this section had been discussed at the Cambridge workshop mentioned earlier. I want to stress again that they all lie at the very forefront at which modern science strives for progress. Quite a few things will be elaborated upon in the future, while much will remain unknown.

It is not clear, for example, what took place still closer to the singularity. There can be hardly any doubt that within a period less than 10^{-45} s after the singularity, both time and space existed as discrete quanta. However, we can only guess what had happened, how, and why.

What existed before the expansion began? Nothing reliable is known about it so far. Certain daring hypotheses could be quoted. In fact, they do not belong to science yet. In a book, one is allowed to fly on the wings of fantasy without serious science at the controls, but it should be a book of a quite different genre than this one.

On the wings of time

Time and again in the life of a creative physicist, especially a theorist, there comes a moment when it seems that the field he or she was studying holds no more problems of interest. People who knew the famous theorist Lev Landau recall the young Landau complaining that like beautiful girls, who are all either engaged or have already married, all problems that deserve attention have all been solved and one can hardly hope to find anything worth the trouble among the leftovers. Of course, Landau himself disproved his jocular remark, having formulated and solved numerous excellent problems. Attractive girls getting married are replaced by younger girls, even more beautiful; likewise, problems already solved are replaced with new, even more fascinating problems.

I remember one such unhappy period in my life: I felt that the problems I was working on held no promise. I mentioned this feeling in a talk with a colleague, a very attractive young woman. The talk shifted to the future of individual celestial bodies. This colleague of mine suggested the interesting problem of calculating the process of the cooling of neutron stars under the exotic conditions of the very remote future of the Universe. In fact, the story that I will outline in the remaining part of this book is based on this joint work, the joint analysis of the work of other scientists on the conditions in the remote future.

The past is studied in order to understand better the present and the future, while the near and distant futures of mankind, the future of intelligent life, depend to a large extent on the future of nature, on the fate of the Earth, the Sun, the Galaxy, and the Universe.

The study of the future of the Universe is principally different from the study of its past. The past has left its traces, and the correctness of our notions can be checked against these traces once we discover them. A picture of the future is no more than extrapolation, which excludes the possibility of verification. Nevertheless, our current foundation of physical and astronomical knowledge is so solid that the remote future of the Universe can be treated with sufficient certainty.

The future depends first of all on whether the expansion will last eternally. First we consider the future of a perpetually expanding homogeneous Universe whose density does not exceed the critical value. What are the processes that unfold in this permanently expanding Universe?

The first of these processes is something nobody doubts anymore: stars will ultimately burn out. The Sun will complete its active evolution in several billion years and turn into an Earth-size white dwarf that will gradually cool down.

Stars more massive than the Sun will have even shorter lifetimes and will turn, depending on their masses, into either neutron stars of only tens of kilometers in diameter, or into black holes.

Finally, a star may come to a catastrophic end, when an explosion completely destroys it. It appears that some stars, known as supernovas, explode in this manner.

Stars less massive than the Sun live longer but, sooner or later, they inevitably turn into cold dwarfs.

At present, new stars are constantly being born from the interstellar medium. The time will come, nevertheless, when the required stores of matter and nuclear energy become exhausted, new stars cease to be born, and old stars turn into cold bodies or black holes.

The stellar phase of the evolution of our Universe will be completed in about 10^{14} years. This is an enormously long period, 10 thousand times the duration from the Big Bang to the present day.

We will look now at the fate of galaxies.

These stellar systems consist of hundreds of billions of stars. It appears likely that the central regions of galaxies contain superheavy black holes which are manifested by the violent processes that astrophysicists observe in galactic nuclei. Sometimes a star in a galaxy gains a very high velocity owing to the gravitational interaction with other stars, leaves the galaxy, and becomes an intergalactic rover. Such events, very rare in this epoch, are essential for the future of galaxies. Stars will gradually leave galaxies whose central parts will contract little by little and finally turn into very compact stellar clusters. The stars of such a cluster will collide, converting into gas, and this gas will mostly fall into the central supermassive black hole, thereby increasing its mass. Stars will also be destroyed by tidal forces when they pass too close to this black hole.

The final stage is a supermassive black hole that has absorbed the remains of the stars in the central part of the galaxy, and the scattering of about 90 per cent of all stars in the outer parts of the galaxy to intergalactic space. The process of galactic decay will be completed in about 10^{19} years when all stars are long dead and actually cease to be stars.

The decisive factor for further processes is the instability of nuclear matter. Namely, it is assumed that the proton is a particle with an extremely long lifetime, but nevertheless an unstable particle. The 'grand unification' theory, which predicts violent processes in the interval from 10^{-34} to 10^{-32} s after the Universe began to expand, also predicts that the proton has to decay (neutrons in compound nuclei, previously assumed to be stable, are also predicted to be unstable). Its mean lifetime is estimated to be approximately 10^{32} years. The final products of proton decay are a positron, a photon, a neutrino, and possibly one or several electron–positron pairs. Even though the search for evidence of proton decay has so far been unsuccessful, very few physicists doubt that some day the discovery will be made.

Nuclear matter will thus decay completely in about 10^{32} years. Even cold stars will disappear. In fact, the decay of nuclear matter will become important in the evolution of the Universe long before that moment. Positrons created in nucleon decay (protons and neutrons are called by the generic term *nucleons*) will be annihilated in collisions with electrons and transform to photons which, together with photons created by the decay of nucleons, will heat up the matter. Only neutrinos freely escape from the star, carrying away about 30 per cent of the entire decay energy.

This decay sustains the temperature of the dead stars and planets at a low level that is nevertheless considerably higher than zero. Thus, white dwarfs cool down during 10^{17} years to a temperature of 5 K but then retain this temperature as long as energy is released by nucleon decays in their cores. Neutron stars cool down to about 100 K in 10^{19} years, after which nucleon decays sustain this temperature.

After 10^{32} years, nuclear matter will decay completely: stars and planets will convert to photons and neutrinos.

The fate of the gas scattered throughout space is different (this gas, left after galaxies perish, may amount to about one per cent of the entire matter in the Universe). The nuclear matter of this gas will undoubtedly also decay in 10^{32} years. However, the positrons created in its decay will be unable to be annihilated with electrons since the extreme rarefication of the gas makes collisions of these particles extremely improbable, and the result is a rarefied electron–positron plasma.

By this time, that is, in 10^{32} years, the Universe will also contain black holes born of massive stars after they die out, and supermassive black holes formed in central regions of galaxies (we will outline their fate a little later).

What is going to happen in the Universe after the nuclear matter completely decays?

The Universe will contain in that very remote future photons, neutrinos, electron–positron plasma, and black holes. The main part of matter will rest with photons and neutrinos. It is into these two types of matter that the ordinary matter decays. The era that starts is radiation-dominated. One must remember, though, that this radiation has considerably cooled down.

As the Universe expands, the mass density of the radiation rapidly decreases because both the particle number density and the energy of each quantum (hence, its mass as well) keep decreasing. In contrast to radiation, the mean density of ordinary matter (represented by electron–positron plasma and black holes) decreases only as a result of the reduction in its concentration in the expanding Universe. Hence, the density of these types of matter diminishes at a slower pace than that of radiation. Consequently, the matter density after 10^{33} years is mostly determined by the mass trapped in black holes. This mass is much greater than the mass of the electron–positron plasma. If the neutrino mass is nonzero, a considerable fraction of the entire mass rests with neutrinos. The radiation-dominated era is replaced by the era of black holes.

However, black holes are not eternal either. As we know, particle creation takes place in the gravitational field close to a black hole; we also know that black holes of stellar and higher mass generate quanta of radiation. This process diminishes the mass of black holes which are gradually converting to photons, neutrinos, and gravitons. This is an extremely slow process. A black hole of, say, 10 solar masses evaporates in 10^{69} years, and a supermassive black hole a billion times more massive evaporates in 10^{96} years. Nevertheless, all black holes will transform with time into radiation, which will again become dominating: a new radiation-dominated era. However, this radiation is incomparably colder than the radiation in the era of matter decay. As we have already mentioned, the radiation density in the expanding Universe decreases at a greater pace than the electron–positron plasma density, so that this plasma will dominate the Universe in about 10^{100} years when it will become virtually the only thing in the world.

At a first glance, the pattern of the evolution of the Universe in the future is very bleak, manifesting only gradual decay, degradation, and scattering.

At the age of 10^{100} years, the Universe will hold almost nothing but electrons and positrons scattered through space at a fantastically low density: the volume per particle equals 10^{185} volumes of the entire Universe visible to us now. Does it mean that all processes will die down, that active motions of the physical forms of matter will be impossible, that no complex systems will be able to form, let alone systems in which reasonable beings would evolve?

This conclusion is not valid. From our current standpoint, all processes in the future will definitely be extremely slow; however, spatial scales will also be quite different from those we observe now. You may remember that when the Universe began to expand and its temperature was in excess of 10^{27} K, matter was being created and violent reactions took place on a time scale of 10^{-34} s and on a spatial scale of 10^{-24} cm. In comparison with such scales and superfast processes, today's events in the Universe, including biological processes, are frightfully slow and spread over enormous distances. The well-known American physicist Freeman Dyson believes that complex forms of motion of matter will be possible in any, no matter how remote, future; intelligent life may even be expected, in very unusual forms, so that 'the pulse of life will beat slower and slower but will never stop'.

In an endless universe stars perish.
Night is immense.
Alone, a chill, black wind
quakes whole heavens with a white fluttering
heaved up from an infant's infinitesimal palms
as soon as he began to know
himself, this world, the stars, centuries,
birth and life and death of everything,
life's end,
the endlessness of heaven.
And no one will ever come to number
these flutterings the world contains –

those shards of light in the abyss.
The night is immense.
And while the child sleeps
New galaxies and men are born. And stars
are born, bloom, and wither
while he still sleeps (he sleeps alive)
cocooned in a quaking web of light
spun out from remote, dead stars.
And what remains there when a man is gone?
This fluttering, this wisp with childish eyes –
scintillation in the icy dark.

Marina Katys
(Translated by Phil Nix and Jim Beall)

Our imagination tries to roam through vistas in the unbelievably distant future. Such time travels are most likely to come across absolutely unpredictable situations. So far we have been contemplating processes implied by reliably established laws but in the future of the Universe there may occur physical conditions that cannot be realized in our experiments (superlow temperatures and densities, etc.) so that the world may be governed by forces and processes that are quite unknown to us. These forces and processes may drastically change the picture.

One of these putative processes is the decay of the vacuum, and its transformation in the expanding Universe into real matter. The vacuum that we called the 'false vacuum' may have already decayed in the epoch of 10^{-34} s mentioned earlier, having produced high-energy particles and antiparticles. This energy corresponded to the temperature of 10^{27} K; the matter density was 10^{74} g/cm^3.

The vacuum of today (colloquially known as empty space) may also carry a certain energy density. As we have seen in the section on the gravitation of empty space, this density is either zero or very low, corresponding to not more than 10^{-28} g/cm^3 or even much lower. This density can hardly be discovered even in astronomical observations. The theory does not exclude the scenario in which the vacuum mass density transforms, some time in the future, into real particles and antiparticles and thus gives rise to new physical processes. The matter created in this transition would certainly be rarefied but nevertheless incomparably denser than 'our matter' spread through the vastly expanded Universe at that moment. This 'phase transition' of the vacuum may prove to be extremely important for the fate of the Universe. In principle this transition may stop the expansion of the Universe and trigger its contraction. The 'supertrue vacuum' produced thereby will possess attractive gravitational properties, in contrast to repulsive gravitation of the 'false vacuum'. Obviously, this reversal of expansion to contraction will thoroughly change the future of the Universe outlined earlier in our story.

I will add one remark. When describing the future of the Universe, I assumed that the mass of neutrinos (of any species) is zero; that is, they behave as radiation. It was also assumed that like photons, these particles have mass only because they always move at the speed of light, being of zero 'rest mass'. But we saw in the chapter on the 'neutrino Universe' that there is a certain probability that neutrinos are massive particles.

This modification may affect the fate of the Universe in two ways.

If the neutrino mass is very small, say, less than a hundred-thousandth of the electron mass, the gravitation created by these particles in the entire Universe will also be very weak, producing no effect on the rate of expansion. However, the mass density of neutrinos will not decrease in the remote future as fast as that of photons but rather as that of ordinary particles, so that the electron–positron plasma will have a permanent slight admixture of neutrinos (and antineutrinos) of nonzero mass.

If, however, it is found that the neutrino mass is close to the predicted upper limit (approximately 0.000 05 of the electron mass), the total mass of these particles in the Universe will be found to be very high, with the mean density exceeding the critical level (10^{-29} g/cm^3), so that the gravitation of neutrinos will at some time stop the expansion of the Universe. This may happen much earlier than the decay of the entire nuclear matter, and even long before all stars are dead. What awaits the Universe in the future in this case is then compression, the destruction of celestial bodies, and ultimate re-creation of superdense, superhot matter going through superviolent physical processes.

You can see that whatever the scenario of the evolution of the Universe, its future is breathtakingly interesting and diversified.

True, the Universe of the remote future is completely different from today's Universe around us – in all contemplated versions. This state is either very much rarefied and cold, or very dense and hot.

This is how things stand. One has to understand it clearly. The evolution of the Universe is continuous: its past was extremely peculiar and quite unlike the present. The future will be equally far from anything that we are currently observing. We must also clearly realize that this future implies nothing fatally inescapable for intelligent life (in the broad sense of this word). The human mind had penetrated numerous secrets of nature and made its laws serve the purposes of mankind.

If we behave sufficiently reasonably and do not let life become extinct on Earth in this epoch of social eruptions (we believe it is achievable and work for this future), it is difficult to predict what heights of scientific power will be reached by mankind in a hundred, a thousand, a million, and even in billions of years. Hopefully, man will learn how to use all the laws of the evolution of the Universe for the benefit of civilization, and how to control these laws. It would be naïve to believe that the Universe has prepared beneficial 'hothouse' conditions for the existence of mankind at all times. 'We should not beg for nature's favors. Our task is to get them ourselves.' These words of a famous naturalist are proud words worthy of mankind. Of course, when work on this scale becomes the order of the day, problems of environmental protection and many other problems will require new approaches. There can be no doubt that the societies of the future will find ways of coping with the problems.

Before putting the last full stop to this story, we have to remind ourselves that any appreciable changes in the Universe (as compared with its present state) will appear in a very long time, both on our day-to-day and astronomical scales, that is, in at least tens, and perhaps thousands, of billions of years. This is much longer than the present age of the Universe, which cannot be greater than the 10–20 billion years since the beginning of expansion.

Conclusion

There can be no conclusion to a story about the subject of this book. The story of an eternally evolving, eternally young Universe rejects the drawing of a line to it. The point we want to especially emphasize is the particularly rapid progress in the science dealing with the Universe – cosmology – that keeps presenting new surprising discoveries and new fundamental knowledge. Now a new generation of young scientists is coming to grips with cosmology's currently 'unsolvable problems'. I wish to see them equipped not only with knowledge but also with inspiration and devotion for their field of research.

References and further reading

Alpher, R. & Herman, R. (1953). *Ann. Rev. Nucl. Sci.*, **2**, 1.

Barrow, J. & Silk, J. (1983). *The Left Hand of Creation*. New York: Basic Books.

Blandford, R. D. & Thorne, K. S. (1979). In *General Relativity: an Einstein centenary survey*, ed. S. W. Hawking and W. Israel, p. 454. Cambridge: Cambridge University Press.

Brown, L. (1976). Letter by W. Pauli to a group of physicists, 4 December 1930. *Phys. Today, Sept. 1976*, p. 23.

Burbridge, G. & Burbridge, M. (1967). *Quasi-Stellar Objects*. San Francisco: W. H. Freeman.

Chandrasekhar, S. (1972). *The Observatory*, **92** (990), 173.

Chandrasekhar, S. (1983). *The Mathematical Theory of Black Holes*. Oxford: Clarendon Press.

Dicke, R. H., Peebles, P. J. E., Roll, P. G. & Wilkington, D. T. (1965). *Ap. J.*, **142**, 414.

Doroshkevich, A. G. & Novikov, I. D. (1964). *Dokl. Akad. Nauk SSSR*, **154**, 809.

Doroshkevich, A. G. & Novikov, I. D. (1978). *Zh. Eksp. Theoret. Fiz.*, **74**, 3.

Doroshkevich, A. G., Zel'dovich Ya.B. & Novikov, I. D. (1965). *Zh. Eksp. Theoret. Fiz.*, **49**, 170.

Field, G. B. & Hitchcock, J. L. (1966a). *Phys. Rev. Lett.*, **16**, 817.

Field, G. B. & Hitchcock, J. L. (1966b). *Ap. J.*, **146**, 1.

Gamov, G. (1948). *Phys. Rev.*, **74**, 505.

Ginzburg, V. L. (1964). *Dokl. Akad. Nauk. SSSR*, **156**, 43.

Gursel, Y., Sandberg, V. D., Novikov, I. D. & Starobinskij, A. A. (1978a). *Phys. Rev. D*, **19**, 413.

Gursel, Y., Sandberg, V. D., Novikov, I. D. & Starobinskij, A. A. (1978b). *Phys. Rev. D*, **20**, 1260.

Hawking, S. W. &Ellis, G. F. R. (1973). *The Large-scale Structure of Space–Time*. Cambridge: Cambridge University Press.

Hawking, S. W. & Israel, W. (ed.) (1987). *Three Hundred Years of Gravitation*. Cambridge: Cambridge University Press.

Kuznecov, B. (1965). *Essay about Einstein*. Moscow: Nauka.

Leavitt, H. S. (1960). In *Source Book in Astronomy*, ed. H. Shapley, p. 188. Cambridge, MA.

Markov, M. A. (1964). *Neutrino*. Moscow: Nauka.

Markov, M. A. & Frolov, V. P. (1970). *T.M.F.*, **3**, 3 (in Russian).

Michell, J. (1784). *Phil. Trans. R. Soc. Lond.*, **74**, 35. Reprinted in Detweiler, S. (1982). *Black Holes*. Selected reprints. Stony Brook, NY: Am. Ass. Phys. Teachers.

Misner, C. W., Thorne, K. S. & Wheeler, J. A. (1973). *Gravitation*. San Francisco: W. H. Freeman.

Newton, I. (1692). Letter to Richard Bentley, 10 December 1692, reprinted in *Gravitation*, C. W. Misner *et al.*, p. 755. San Francisco: W. H. Freeman.

Novikov, I. D. (1983). *Evolution of the Universe*. Cambridge: Cambridge University Press.

Oppenheimer, J. R. & Snyder, H. (1939). *Phys. Rev.*, **56**, 455.

Penzias, A. A. (1979). The origin of the elements. Nobel lecture, 8 December 1978. *Rev. Mod. Phys.*, **51**, 430.

Penzias, A. A. & Wilson, R. W. (1965). *Ap. J.*, **142**, 419.

Pontecorvo, B. (1983). *Priroda, 1983* (1), 43 (in Russian).

Price, R. H. (1972). *Phys. Rev.*, **95**, 2419.

Sciama, D. W. (1969). *The Physical Foundation of General Relativity*. London: Heinemann.

Shapiro, S. L. & Teukolsky, S. A. (1983). *Black Holes, White Dwarfs, and Neutron Stars*. New York: Wiley Interscience.

Silk: J. (1980). *The Big Bang*. San Francisco: W. H. Freeman.

Thorne, K. S. (1974). *Scient. Amer.*, **231** (6), 32.

Weinberg, S. (1977). *The First Three Minutes*. New York: Basic Books.

Wheeler, J. A. (1960). *Neutrinos, Gravitation and Geometry*. Rendiconti della Scuola Internazionale di Fisica 'Enrico Fermi', Corso XI, Bologna, p. 67.

Wilson, R. W. (1979). The cosmic microwave background radiation. Nobel lecture, 8 December 1978. *Rev. Mod. Phys.*, **51**, 440.

Wolf, J. A. (1967). *Spaces of Constant Curvature*. New York: McGraw-Hill.

Zel'dovich, Ya. B. (1971). *Zh. Eksp. Theoret. Fiz. Pisma*, **14**, 270.